Trap Responses of Flying Insects

Trap Responses of Flying Insects

The Influence of Trap Design on Capture Efficiency

R.C. Muirhead-Thomson

Formerly Leverhulme and MRC Fellow, Royal Holloway and Bedford New College, (University of London), Egham, Surrey, UK

ACADEMIC PRESS
Harcourt Brace Jovanovich, Publishers
London San Diego   New York
Boston   Sydney   Tokyo   Toronto

ACADEMIC PRESS LIMITED
24/28 Oval Road
LONDON NW1 7DX

United States Edition published by
ACADEMIC PRESS INC.
San Diego, CA 92101

British Library Cataloguing in Publication Data
Trap responses of flying insects.
   1. Insects. Behaviour
   595.7051

   ISBN 0-12-509755-7

Printed in Great Britain by St Edmundsbury Press Limited,
Bury St Edmunds, Suffolk
Typeset by Setrite Typesetters Ltd.

Contents

Contents

Introduction

The use of capture, trapping and other sampling methods plays an essential part in all studies on the ecology and behaviour of insects in the field. Under certain favourable conditions a great deal of information about some aspects of insect activity can be gained from field observation alone. But there are obvious limitations to this when investigating other aspects of behaviour, or when studying insects which are mainly active at night. According to the different insects involved, and to the nature of the investigation, these capture and sampling methods have evolved along many different lines and proliferated into a multitude of different designs. Some of these trap designs, such as the light traps used for moths and the light traps used for mosquitoes, have undergone comparatively little alteration over the years, but in general the design and operation of most trapping systems are constantly being revised, improved and modified in the light of experience.

Against this constantly changing background any idea of laying down 'guide lines' or 'manuals of instruction' would be unrealistic. Methods developed for a particular species or for a particular purpose very often prove unsuitable for other, even closely-related species, or other environments. In addition, many research workers are reluctant to accept without question trapping or sampling methods which have been developed elsewhere, without first imprinting their own individuality on the design by modifications or improvements according to local requirements. In many cases these modifications are given names, either that of the research worker himself, the institute he is associated with, or the geographical location of the investigation. Alternatively, the design may carry a simple description, for example the biconical trap, or carry a code number, such as R3. Examples of this proliferating nomenclature of trap designs will crop up repeatedly throughout this book.

Not surprisingly, the bulk of the literature on trapping flying insects comes from the vast accumulation of knowledge about insects of economic importance, i.e. applied entomology, but in recent years significant contributions have come from work on faunal surveys, not necessarily concerned with insects as pests, but rather with the general ecology of endangered species in tropical environments at risk. Progress in that field has sometimes opened avenues of research still unexplored by applied entomologists.

In the field of applied entomology, the two main disciplines involved, namely agricultural and forestry, and medical and veterinary, continue to pursue rather separate and independent courses. Rapid advances in both of these fields have led to increasing specialization, making the task of keeping up with one's own subject so demanding that less and less time and energy can be spared for keeping up with allied subjects. This is well illustrated by two fields of very intense activity, namely the biology and control of tsetse flies with particular reference to insect response to trapping systems, and the even more explosive field of sex-attractants or pheromones of insects of agricultural importance, and the increasingly critical and penetrating research on insect response to pheromone-baited traps associated with this. There are few common meeting grounds for the specialists in these two divergent disciplines. Nevertheless, a striking exception is provided by the fact that increasing emphasis on animal odour-baited traps and on odour ingredients in tsetse research, for example, is inevitably directing research efforts on insect responses into fundamentally similar channels to those being explored in the field of sex pheromones, and the response of insects to different pheromone ingredients.

Traditional separation of agricultural from medical entomology has also in some cases resulted in a kind of language barrier. For example, to plant pest specialists, black flies or blackflies are plant sucking aphids of the bug order Homoptera. To the medical entomologist the same term applies to small two-winged blood-sucking flies (*Simulium*) of a quite different order, Diptera. In trap nomenclature too, the same term may mean different things to different people. To the agricultural and forestry entomologist 'window traps' are vertical sheets of glass or transparent perspex used to intercept the flight of bark beetles and their allies: to the medical entomologist 'window traps' are one-way traps in the form of net cages fixed over the windows of human or animal habitation, either to trap — through a funnel — mosquitoes leaving the habitation after feeding or resting indoors, or in reversed form, to intercept mosquitoes entering the habitation from outside.

In view of this dichotomy, there would appear to be a real need to attempt an overall review within the compass of one book. The increasing specialization would suggest that logically this can only be achieved by a multi-author symposial type of publication, in which the contributors are specialists actively engaged in one or other of the many disciplines. The obvious advantage of that type of publication in providing up-to-date expertise is offset by the fact that few of the contributors would have sufficient space, or opportunity, to extend the scope of their review outside their own subject to any extent. The need for an objective overall review can still perhaps best be met on a single author basis, and such an approach may also offer the best opportunity for continuity of text and theme when passing from one specialized subject to another.

In view of the enormous amount of literature covering this extended field, selection has to be exercised in order to avoid the text from becoming simply a 'summary of summaries'. This selection of material is determined by the title or theme of this book, and consequently most space will be devoted to published work in which the experimental approach predominates, that is, with regard to the various problems encountered in insect response to traps and to trap design, and to attractants whether visual or olfactory. Some of these research projects seem to be particularly worthy of detailed description and assessment, and it is hoped that the additional space they merit may assist researchers in one field to appreciate progress in other unfamiliar fields from which their interest has been discouraged by the sheer mass of highly specialized reports and papers.

Selection of material inevitably means less space for other work, less relevant to the theme of this book, but nevertheless sufficiently important not to be ignored or overlooked. Undoubtedly, important contributions may inadvertently have been omitted, and in some cases overlooked. This is to be regretted, and must be attributed to the fact that the author falls far short of omniscience in undertaking such a formidable task.

Acknowledgements

The preparation of this book was greatly assisted by generous donations from the Carnegie Trust for the Universities of Scotland (Professor J.T. Coppock), and from the Shell Research Laboratory, Sittingbourne, (Dr N.R. McFarlane). In compiling as impartial a review as possible I must make a comprehensive acknowledgement to all the research workers whose publications I have freely quoted, and whose figures have been used to illustrate the text. In the case of unpublished material – still in thesis form – I am particularly indebted to Dr W.J. McGeachie and Dr P.A. Gaydecki of the Cranfield Institute of Technology for the use of their research material on the reactions of moths to light traps. Because of the great significance of these contributions they have been reviewed as fully as is possible within the confines of this book.

Of the many colleagues-through-correspondence who have been so helpful in providing material and up-to-date information I am most grateful to the Australian researchers to Dr Keith Wardhaugh and Dr V.A. Drake of CSIRO Canberra, to Dr G.H.L. Rothschild, Director of the Bureau of Rural Resources, Canberra, and to Professor W. Danthanarayana of the University of New England, Australia. Professor S. Mukhopadhyay of the Bidhan Chandra Krishi Viswavidyalaya Plant Virus Research Centre, West Bengal has provided valuable information about the many developments in India highly relevant to the theme of this book. Among other helpful correspondents I am indebted to Professor M.J. Samways of the University of Natal, to Professor W.L. Roelofs of the New York State Agricultural Experimental Station, and to Dr G.M. Tatchell of Rothamsted Experimental Station.

In trying to achieve as accurate a review as possible of the many publications quoted, all text and diagram measurements dealing with length or area have been retained exactly as in the original, and no attempt has been made to impose a standardized metric equivalent, unless the author of that report has done so. Transforming these Imperial measurements automatically into their metric equivalent – which in many cases would involve five figures and three decimal places – would only give a confused presentation of the original clear-cut experimental layout. This could also apply to the alternative practice of following the original Imperial data by the metric equivalent in parentheses. This again could only lead to a proliferation of figures sufficient to detract from each author's original presentation.

Chapter 1

Light Traps

1.1 Introduction and background

The use of artificial light to attract and trap moths and other nocturnal insects has long been practised by insect collectors in general, and by applied entomologists in particular. From its simple beginning as an electric bulb or a kerosene lamp in front of a white sheet, the development of light traps has progressed rapidly to the more sophisticated and automatic

1

models now available. One of the great advantages of light traps, especially for night-flying Lepidoptera, is that no other trapping method has proved so consistently successful in capturing larger numbers or a greater variety of species. For example, in an 18 month light trap survey in Queensland, Australia, there was a total catch of 750 000 moths, of which 339 000 were noctuids, composed of over 300 species (Persson, 1976). In a survey at Mugugu, near Nairobi, Kenya, up to 49 000 moths have been taken in one trap on one night, at a time when other methods of trapping, using different techniques, only succeeded in capturing a single moth (Brown *et al.*, 1969). In fact, light trap captures of some pest species have been so strikingly high as to lead to attempted control of such pests as cotton bollworm and tobacco hornworm by means of a network of traps alone, with the object of producing a significant reduction in population (Hartstack *et al.*, 1968). At the other extreme, reports from such widely separated regions as Australia, Europe and the USA have shown that such a notorious and widespread pest as the oriental fruit moth is rarely taken in light traps, even in those providing a range of light sources (Rothschild, 1974).

Much of the critical work on light traps has been carried out on moth species whose larvae are important agricultural or forestry pests. But light traps are increasingly being used for a variety of plant pests such as plant bugs and aphids which are mainly important as vectors of plant virus diseases. This is a development which has been especially marked in India, which has a history of light trapping dating back to the first decade of the century (Nath and Banerjee, 1985; Pawar *et al.*, 1985).

In the course of these and other investigations it has become evident that, in addition to moths, a wide range of insect orders, genera and species are regularly attracted to light traps. For example, in one series of studies in Africa (Bowden and Church, 1973) practically all main suborders of insects were taken, including a wide range of beetles, wasps, bugs and grasshoppers. This wealth of species, and sheer volume of catch, could in fact prove an embarrassment to the specialist concentrating on the fluctuations of a few key species only.

Nevertheless, for a subject which seems to have been so exhaustively studied in so many countries, the main impression is that the whole subject of insect capture by light trap and insect response to these traps is one that needs continual re-appraisal and assessment. It is only comparatively recently for example that Australian workers, by making a simple modification to the conventional design of light trap, were able to capture successfully the migrant locust, never previously recorded by this capture technique (Farrow, 1974, 1977). No one can deny the success of light traps in capturing nocturnal insects. But the questions remain: what relation does the light trap catch bear to the total population exposed, and in what way are catches influenced by such imponderables as trap design and efficiency, nocturnal insect behaviour, changes in insect flight density, or variations in flight path? If we add to these variable factors the regular rhythmic changes in night-time illumination in the course of each lunar cycle, we are presented with an ecological challenge sufficient to test the ingenuity of a whole generation of dedicated research enthusiasts.

As in so many other fields of applied entomology, studies on agricultural and forestry pests have tended to follow lines of investigation quite independent of those pursued by entomologists concerned with insect pests of man and domestic animals, the vectors of human and animal disease. In that field, mosquitoes are the predominating night-flying winged insects, along with other small two-winged flies such as biting midges (Culicoides) and sand flies (Phlebotomus). The use of light traps to capture nocturnal mosquitoes and their allies also has a very long history, and continues to be one of the main standard methods for monitoring populations of mosquito vectors of disease, particularly those concerned with transmission of viral disease. The fact that these two main branches of applied entomology have continued to pursue independent courses, even on what appear to be basic insect capture techniques, may be associated in some way with their different origins. The systematic use of light traps for the study of nocturnal moths, including major pest species, originated in the classic studies at Rothamsted Experimental Station, Harpenden, England from the 1930s onwards, from whence its influence subsequently extended to other centuries of the Old World, mainly Africa and India. In contrast, the use of light traps for mosquitoes has long been preeminently an American concern and this has gradually spread to those other parts of the world of concern to American-based or American-influenced teams of research workers.

In the following sections it is hoped to emphasize, not so much the differences in approach adopted by these two different disciplines of applied entomology, but the existence of so many problems common to both which are still unresolved, and which urgently demand closer liaison. In this connection it is encouraging to note that recent studies on trapping night-flying insects in Australia have encompassed both moths and mosquitoes in their programmes (Danthanarayana, 1976). Although the basic capture technique in that study was the use of suction traps rather than light traps, the role of moonlight and moon phase was a major objective of that work, a factor of prime importance in the intepretation of light trap capture data.

1.2 Basic components and design of light traps

1.2.1 Light traps for agricultural pests

Light traps designed for capturing night-flying moths, especially pest species of economic importance in agriculture and forestry, consist essentially of three elements. An electric bulb as attractant, a funnel to direct attracted insects into the third element, a container or collecting bag. Within that framework, the size, composition and construction of the different parts have been modified by different workers according to their special requirements, or different environmental conditions.

The source of light attractant is the single most important element determining the range of insects captured (Vaishampayan, 1985a,b). These lights may be either (a) incandescent tungsten bulbs which emit energy in the long-wave radiation (green−yellow−orange), (b) mercury vapour lamps which emit energy characterized by both short-wave and long-wave radiation, or (c) ultraviolet or 'black light', dominated by violet and blue short-wave radiations. Different groups of agricultural pests may react in different ways to these light sources. For example, the mercury vapour lamp is attractive to a wide range of insect pests, while the ultra-violet attracts the majority of noctuids and Coleoptera, and the incandescent lamps are particularly attractive to leaf hoppers.

The incandescent tungsten-filled, 200 W bulb is the light source used in the classic Rothamsted Experimental Station − or Williams − light trap, which has been in almost continuous use since it was first designed and tested under various conditions in the 1930s (Figure 1.1A,B) (Williams, 1936; Taylor and Brown, 1972). In another design, the Robinson trap, originally designed in the UK and subsequently widely used in light trap studies in Africa and North America, the mercury vapour bulb is the light attractant (Figure 1.2). UV or 'black-light' bulbs have been the main attractant element in light trap studies on mosquitoes and other night flying biting insects.

The Rothamsted light trap has also been widely used in India where it has undergone modifications and improvements in order to deal with specific requirements and simplify its handling. One of these, the Pilani trap (Figure 1.3) was found to have certain advantages, especially when fitted with a UV bulb (Kundu, 1985). Another modification, the Chinsura trap (Mukhopadhyay *et al.*, 1985), was developed specifically to monitor the rice green leafhopper, *Nephotettix virescens*.

In still other cases, the light trap has been designed to monitor a range of insect pests showing wide size differences. One of these, developed in India, the JNKVV trap, named after the research institute involved, the

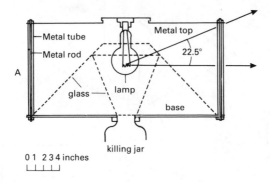

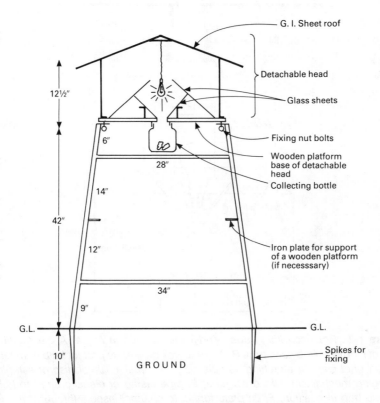

Figure 1.1 *A. Conventional Rothamsted light trap (after Taylor and Brown, 1972).*

       B. Model of Rothamsted light trap (of conduit pipe frame) raised on support, as currently used in India (after Kundu, 1985).

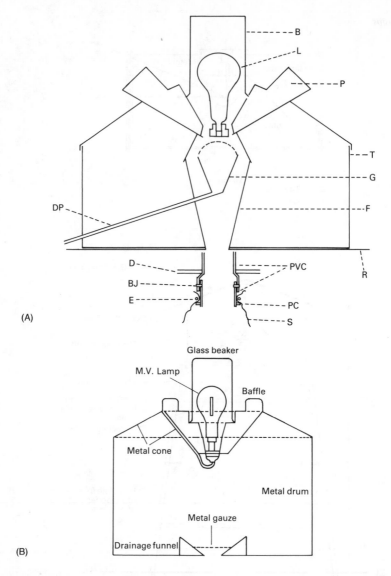

Figure 1.2 *Robinson light traps. Modifications used in Africa. A, Sectional view of modified Robinson trap. B, 2 Litre Pyrex beaker (to protect bulb from rain); BJ, bayonet joint for attaching sample bag; D, circular Tufnol disc of sample-changing mechanism; DP, drain pipe; E, pipe-clamp or eleastic band to hold sample bag in position; F, tin plate funnel to channel insects through trap; G, gauze-capped funnel for rain drainage; L, 125 W MB/U mercury-vapour bulb; P, vertical baffle-plates (four of these); PC, plastic ring glued to PVC pipe; PVC, lengths of PVC drain pipe; R, roof of sheet-iron box; S, sample bag (twelve of these); T, body of Robinson trap. After Siddorn and Brown, 1971, and B. (Muguga) after Taylor and Brown (1972).*

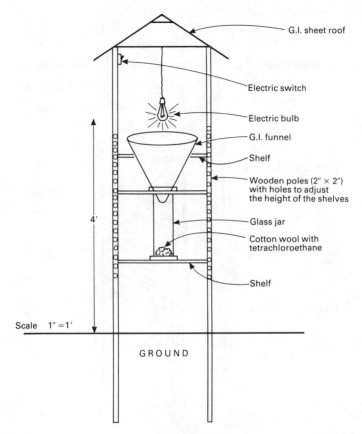

G.I. sheet roof

Electric switch

Electric bulb

G.I. funnel

Shelf

Wooden poles (2″ × 2″)
with holes to adjust
the height of the shelves

Glass jar

Cotton wool with
tetrachloroethane

Shelf

4′

Scale   1″ = 1′

GROUND

Figure 1.3 *Pilani light trap (wooden body), modification of Rothamsted trap currently used in India (Kundu, 1985).*

J.N. Krishi Vishwa Vidyalaya (Figure 1.4), provides for the automatic sieving and grading of insects into four categories, allowing the separation of microsize insects, such as aphids and leafhoppers, from crickets, medium-sized Lepidoptera and large moths (Vaishampayan, 1985a,b).

In addition to the light source and angle of direction, other construction features may determine the effectiveness of different designs. Different methods of retaining trapped insects may be determined by whether trap catches are high or whether low overnight catches are the norm. The whole night collection may be retained in a single collecting chamber, which is provided with a suitable asphyxiant such as the insecticide DDVP. At the other extreme it may be necessary to segregate hourly catches throughout the hours of darkness, either as an essential part of studies on

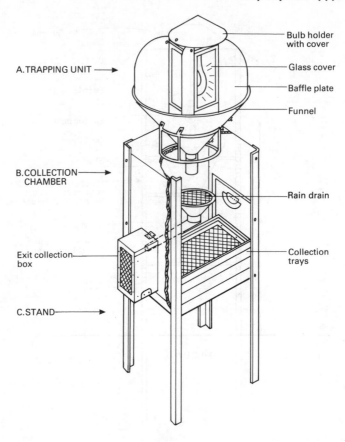

A.TRAPPING UNIT ——▶

Bulb holder
with cover

Glass cover

Baffle plate

Funnel

B.COLLECTION ——▶
  CHAMBER

Rain drain

Exit collection ——
box

Collection
trays

C.STAND ——▶

Cross-Section (Lateral View)

Figure 1.4 *The JNKVV light trap (SM 84 model) for segregating insect pests (after Vaishampayan, 1985b).*

nightly patterns of insect flight or density, or simply because unusually high numbers of captured insects necessitate hourly clearing of the collecting bags. The smooth operation of such hourly segregated trap catches is only possible through the design of sophisticated electronically-controlled equipment, requiring a high degree of technical skill for operation and maintenance.

In the development and modification of various light trap designs, the need for improved trap effectiveness has not always been the dominating factor; in many cases it has been just as important to consider such features as cost, ease of transport and maintenance.

1.2.2 Light traps for mosquitoes and other night-biting Diptera

The possible use of light traps for monitoring populations of nocturnal mosquitoes — as distinct from several important species which are active by day rather than night — was early exploited by public health entomologists in the USA. Pride of place must go to the New Jersey light trap, designed and made widely operative over 50 years ago (Mulhern, 1932, 1942) and still used as a standard technique in mosquito-borne viral surveillance. Its continued successful contribution over the years merited a reprint of the original paper on its 50th anniversary by the American Mosquito Control Association (Mulhern, 1985).

The distinctive features of this trap, in which a standard design was established in the 1930s, is the mains-powered light bulb as attractant, combined with an electric fan to draw attracted insects into a container (Figure 1.5). The light source is a 25 W 110 V white-frosted bulb. A mesh screen across the trap opening excludes moths and other bulkier insects, allowing only mosquitoes, biting midges and sandflies, and similar small Diptera access to the trap. The electric fan has a capacity of $360-400\,\text{ft}^3/$ min, with the air blast directed downwards into the container where the insects are killed by calcium cyanide.

The New Jersey trap not only continues to be the standard mosquito trapping technique in the encephalitis surveillance programme in North America (Reisen and Pfuntner, 1987; Easton, 1987) but has been the favoured capture method used by all US public health entomologists working overseas in such areas as the Caribbean, Central and South America, Korea, Viet-Nam, the Philippines, as well as various Middle Eastern countries. In many of these surveys an additional smaller model of light trap has frequently been used (Sudia and Chamberlain, 1962). This is the CDC miniature light trap — named after the Communicable Disease Centre, Atlanta — which is battery-operated, usually by 12 V motor cycle battery, and hence has the advantage of being usable away from main power supplies, as well as being easily portable. Originally both types of trap used incandescent bulbs, but in recent years UV has also been used successfully; in the case of the CDC trap, the attractant value is usually further enhanced by means of dry ice (Sexton *et al.*, 1986).

In addition to their role in monitoring mosquito populations, light traps have also played an important part in experimental studies on mosquito

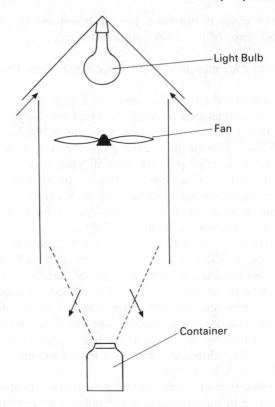

Figure 1.5 *The New Jersey light trap for mosquitoes (after Mulhern, 1942, 1985).*

flight range and dispersal. This line of enquiry has been actively developed in the USA and in Canada.

One of the most noteworthy of these field experiments was carried out in Manitoba, Canada (Brust, 1980) on floodwater-breeding species of *Aedes*. In this type of field trial, huge numbers of larvae required to produce the very high numbers of adults necessary — up to half a million at a time — are collected and raised, marked with fluorescent powder and released from a central point. The main concern here, however, is the role of the light traps. In the first test in 1976, 29 New Jersey light traps equipped with 100 W incandescent bulbs were located within an 11 km radius of the central release point. Traps were operated continuously, the catch being collected daily for three weeks, and then every other day. In the following year (1977), 1.25 million adult *Aedes* were released over a

4-d period, and their dispersal checked by means of 38 light traps within a radius of 8 km.

Another main target of such mark−release−recapture experiments involving light traps has been *Culex tarsalis*, a primary vector of Western equine encephalitis and St Louis encephalitis in the western USA. In one of the main trials carried out in California (Nelson *et al.*, 1978) equal numbers of males and females, marked with four different colours of fluorescent dusts according to release points, were released at the rate of 1000 per day from the centre of each zone. In this case the light trap used was the battery-operated CDC design, supplemented by CO_2. Collections were made on ten consecutive nights after release in 24 of these. CDC light traps were also used in essentially the same manner in studies in California involving the release of genetically altered male mosquitoes (Asman *et al.*, 1979; Nelson and Milby, 1980).

1.3 Experimental studies on light trap performance

1.3.1 Introduction to problems of evaluation

Light traps have proved to be an extremely productive capture technique for two distinctly different groups of night-flying insects, namely moths and mosquitoes. Systematic trapping has provided an immense amount of data relevant to insect distribution, abundance and flight patterns. There are different ways of interpreting and assessing this capture data. The first is to compare the performance of different light trap designs under comparable conditions, and to examine the performance of the same trap under a range of topographical and climatic conditions. The second approach is to compare capture data from light traps with capture data for the same species or groups of insects, obtained by alternative capture techniques which do not involve a light attractant. In the case of mosquitoes and other nocturnal biting Diptera this second approach has proved the most obvious and instructive line of investigation because of the wealth of alternative capture and sampling techniques available. In most cases, mosquito capture data from light traps can be compared with simultaneous captures on human and animal bait using suction traps, dry ice-baited traps and collections of resting insects during the inactive daytime period. Such comparisons show that, for example, some species of biting insects, rarely taken in light traps, may be captured in high numbers by one or other of the alternative capture methods, and vice versa. These alternative capture methods can also confirm that, in some cases, when closely related species are consistently recorded as existing at widely different population levels, the disparity does not reflect a wide difference in abundance, but can readily be attributed to differences in trap response to the species involved.

The interpretation of light trap captures of moths has long been rendered difficult by the fact that, for most species, no other sampling technique was available. It is only in comparatively recent years that effective alternative methods for sampling moths in flight, i.e. pheromone traps and suction traps, have contributed so much to the interpretation of light trap capture data.

1.3.2 Studies on light traps in Africa

Of the many studies which have been carried out on comparison of different light traps, and of factors affecting trap performance under different conditions, work on the African army worm, *Spodoptera exempta*, in Kenya has been particularly instructive. Seasonal outbreaks of African army worm have long been experienced in East Africa and parts of southern Africa, and this led to the initiation of intensive light trap studies from 1963 onwards, with the object of measuring changes in the distribution of populations (Brown *et al.*, 1969; Siddorn and Brown, 1971). Because of the huge number of moths liable to be trapped in the course of a night — up to 49 000 on one occasion — it was necessary to subdivide the night catch by means of an automatic segregating mechanism. This practice was also in accord with the added objective of defining nightly flight patterns based on hourly records. Each collecting container of the trap was provided with a strip of insecticide so that captured moths did not batter themselves to pieces, or trapped beetles chew their way out of the polythene bags.

Initially, the light, which could be operated from AC mains or from a small petrol electricity generator, was left on all night, but this practice had to be abandoned for two reasons. First, on nights of exceptionally high moth abundance, each bag could overflow within the hour and jam the mechanism, and secondly, it was noted that on nights of high moth abundance, a swarm of flying moths tended to gather round the light when it was first switched on; the swarm tended to remain there indefinitely throughout the night, with moths only entering the trap a few at a time. Hence moths attracted to the light trap early in the night might not actually enter the trap until later in the night, rendering the trap results invalid. This was dealt with by means of a multiset switch which enabled the light to be switched off for a quarter hour period each hour. This had the effect of disrupting the swarm and periodically releasing moths from being held indefinitely in the vicinity of the trap.

The light trap studies in Kenya, originally directed towards the pest species, *Spodoptera exempta*, eventually extended in scope to provide valuable information about many other species and groups of moths. The experimental layout in one of these studies on trap performance is particularly instructive (Taylor and Brown, 1972). It involved a comparison of two structurally different trap designs; one of them was the standard Rothamsted (Williams) trap with tungsten filament bulb (Figure 1.1a) while the other was a modification of the Robinson trap with mercury vapour lamp (Figure 1.2b). The former had a 60 cm square roof over the lamp, confining the illumination to a horizontal beam which diverged about 22.5° above the horizontal. The latter design had no roof, and the light shone upwards from the hemisphere of the bulb, which was protected from rain by means of an inverted 2-l beaker.

In one experiment, the inverted beaker was painted with aluminium in three different ways (Figure 1.6): (a) leaving only 10% of the surface transparent, i.e. 90% obscured; (b) with 25% of the surface transparent, i.e. 75% obscured; and (c) leaving only the base, i.e. the uppermost part, of the inverted beaker transparent to light, so that the light shone upwards only. Each of the five trap/light combinations were exposed in a rotating pattern which changed each night. The design of the trap tests allowed a comparison to be made of catches according to direction of illumination. The results indicated that different taxa of moths tend to fly at different heights in relation to the traps, which were usually set at 60 cm above

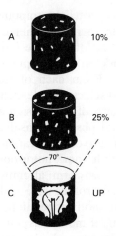

Figure 1.6 *Inverted beakers used as lamp covers in light experiments with different degrees of exposure (after Taylor and Brown, 1972).*

ground. Pyralids, for example, were more abundant in the Rothamsted trap with its near horizontal light beam, suggesting that this group flies nearer the ground than the progressively higher flying noctuids and sphingids, which recorded highest catches in trap combinations with vertical illumination.

For each of these traps, two light sources were also compared; in one series the Rothamsted trap had the normal 200 W clear glass tungsten bulb, while in the second series the trap was fitted with a 125 W frosted glass mercury vapour lamp. Two different lamps were also used in the Robinson-type traps. In each experiment one was fitted with a 125 W frosted glass mercury vapour bulb, and the other with a 125 W frosted glass, mixed tungsten filament and mercury vapour bulb.

These field trials are instructive in another way in that they draw attention to the number of trap designs and operational factors which enter into such comparison experiments. These experiments involved eight different trap/light combinations in two different experiments in five different sites. An additional variable was created by the fact that as all the traps, except the Rothamsted design with its metal cover, were visible simultaneously from above, it seems certain that the high moth catches recorded — greater than could be accounted for by local origin — must have been due to immigrant moths approaching the site downwind.

1.3.3 Experiments in North America on boll worm and cabbage looper moth

The USA has a long history of light trapping nocturnal insects, going back over 100 years. As in so many other cases, studies on light trap problems involving moths and other large insects have been carried out quite independently of work on light trapping of mosquitoes and other small night-biting Diptera. From each of these disciplines, it will be sufficient for the moment to select a single example illustrating the experimental approach to problems of trap design and performance.

In the field of agricultural pests, a considerable amount of work has been done on the systematic use of light trapping as a means of estimating total insect populations. With some notorious pests such as the cotton boll worm and the tobacco hornworm, attempts have been made to use a network of such traps as a control measure with the aim of trapping out a significant proportion of the moth population. It should be mentioned at this point that light trapping of moths has not always met with success in North American species. The spruce budworm, *Choristoneura fumiferana* has proved refractory to light trapping, as also has the mainly diurnal Gypsy moth, *Lymantria dispar*.

Many observers have noted that of the huge number of night-flying insects attracted to such traps most appear to be circling the traps and settling on surrounding vegetation. The vital question of what proportion of these are actually trapped was examined critically in an experiment whose design opened up a completely new approach to this problem (Hartstack *et al.*, 1968). Around a light trap in which the attractant was a 15 W blacklight bulb, shallow pans 14 × 14 cm × 3 mm deep, filled with water, were placed in a grid system with the object of trapping those insects which alight on the ground. These pans, 114 in total, were arranged in six concentric circles round the trap at radii from three to 48 feet (Figure 1.7). The number of pans in each circle ranged from 12 in the innermost circle to 48 in the outer one. The pan traps were filled with 1 part diesel oil to 3 parts of water. In preliminary tests, with the filled pans

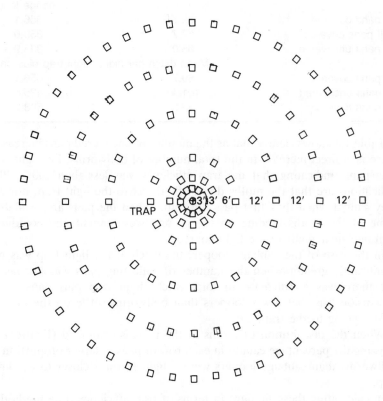

Figure 1.7 *Layout of pans surrounding light trap in N. American experiments. (after Hartstack* et al., *1968).*

uncovered, but with no light, only one or two moths were taken in the pans. Three experimental treatments were used:

(i) The light trap in the test area was used alone, all pans being covered.
(ii) The light trap was operated, with half the number of pans — odd and even — exposed.
(iii) The light trap was operated with all the pans uncovered.

The results were sufficiently striking to merit tabulation in some detail, in which the figures for the bollworm, *Heliothis* are compared with those for the cabbage looper moth, *Trichoplusia* (Table 1.1).

Table 1.1

	Mean catch per night, light trap	
	Bollworm	Cabbage looper
All pans covered	89.7	356.1
Half pans covered	67.7	330.0
All pans uncovered	65.0	336.9
	Mean catch per night, Light trap plus pans	
All pans covered	89.7	356.1
All pans uncovered	101.4	425.6
All pans covered	122.5	508.2

Table 1.1 shows clearly that as the number of uncovered pans increases, there is a direct increase in the total number of bollworms, i.e. light trap plus pans, indicating that the trap efficiency was less than 100%. The indications are that the moths do not fly direct to the light trap, nor are they caught when they first pass the trap. When the pans are available, some of the moths landing on the ground were caught and could not continue their flight to the light source.

In the case of the cabbage looper, the catch at the light trap was not significantly greater when the number of collecting pans was increased, but there was an increase in total catch (light trap plus pans). The conclusion was that fewer loopers than bollworms settle on the ground before going to the trap.

When the distribution of moths in the pans is examined (Figure 1.8) showing the percentage caught in each row of pans, a larger proportion of bollworms than cabbage loopers were caught in traps closer to the light trap.

In calculating these findings in terms of trap efficiency, one method is to assume that each pan represents a certain number of square feet of area around the trap. By multiplying this factor by the catch in each pan, an estimate can be made of the total population that landed in the light trap area, assuming that each moth lands only once. This gives an estimated population present, but not caught in the trap, as 771.19. To this is added the 65 for the trap catch — with all pans uncovered — to give the total moths present as 836.19, of which only 10.73% were taken in the light trap. Put in another way, the 'trap efficiency' of the light trap proved to be little more than 10% in the case of bollworm.

If it is assumed that each moth makes more than one landing in the sample area, then the calculation becomes more involved, and subject to error. But it does suggest that for the bollworm moth, trap efficiency is higher — around 50% — under these conditions whatever method of calculation is used.

In the case of the cabbage looper, the first analysis — assuming one landing only — gives a trap efficiency of 8.21%, which is increased to 38.4% by the second method of calculating in which more than one landing is postulated.

1.3.4 Experimental studies on macro- and micro-Lepidoptera in England

After a lapse of nearly 20 years, the full potential of the experimental approach described above has been revived and further developed in a comprehensive study in England, embracing a wide range of moth species (McGeachie, 1987, 1988). This used the same basic principle of light traps surrounded by concentric rings of water pan traps. In the analysis a distinction was made according to size between the Macrolepidoptera, such as noctuids, geometrids, arctiids and *Notodontia*, and the Micro-lepidoptera, e.g. pyralids and tortricids. A further more detailed analysis was made of the reactions of the three most abundant species, the noctuid *Apamea anceps* (Large nutmeg), the hepialid *Hepialus lupulinus* (Common swift) and the pyralid *Chrysoteuchia colmella*.

As this investigation involved not only the constant illumination factor provided by the light trap, but also three variable factors, namely temperature, wind velocity and wind direction, and also had to take into account the role of the unsampled areas between water pans, a very exhaustive computer analysis was essential for evaluating the great mass of data provided.

(a) Experimental layout of water traps

Two sites, experimental and control, were set up 200 m apart, each centred on a Robinson type light trap (125 W MV lamp), the trap being slightly modified by replacing the base with a plastic container painted matt black, so designed to enable placement of water traps nearer the

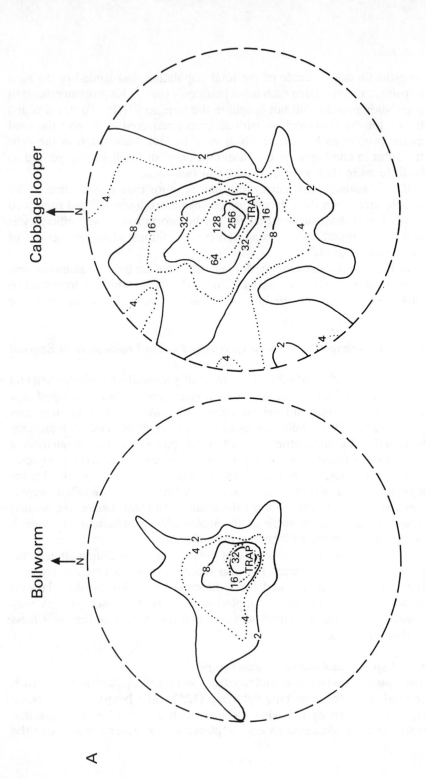

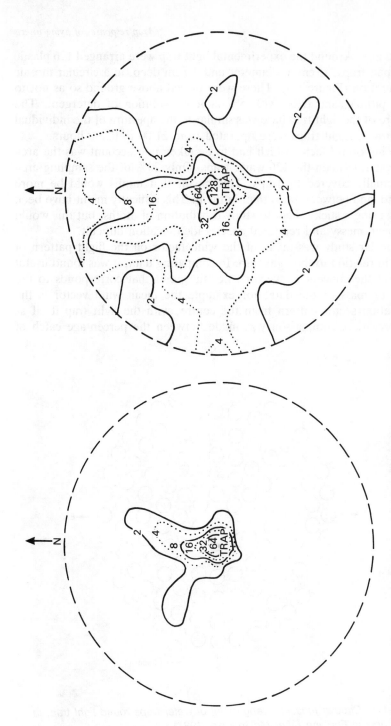

Figure 1.8 *Distribution of bollworm moths and cabbage looper moths in pans round light source. A. With all pans open; B. with half pans open (after Hartstack et al., 1968).*

light source. Around the experimental light trap were arranged 126 plastic water pan traps, 15 cm in diameter and 3.7 cm deep, in a circular mosaic configuration (Figure 1.9). These were placed above ground so as not to act as pitfalls, and filled with 5% aqueous solution of detergent. The aperture of the light trap base was similar to the aperture of the individual water traps. Light traps were operated from 21.00 to 02.00 hours.

One important factor which had to be taken into account was the area unsampled between the 126 water traps. Only 20% of the sampling area was actually covered by traps. This meant that moths would be more likely to land between traps than in them. This difficulty might have been met by using a sticky sheet to study distribution of moths, but this would have been messy and rendered species identification difficult.

Particular study was made of the wind factor on the flight pattern of moths in relation to the light trap. In comparing data, it was found useful to adopt the concept of 'vector', i.e. the point that corresponds to the centre of mass of the data; for example the mean wind vector is the distribution sector pattern from the centre, with the light trap itself as zero vector. A comparison was made between the percentage catch of

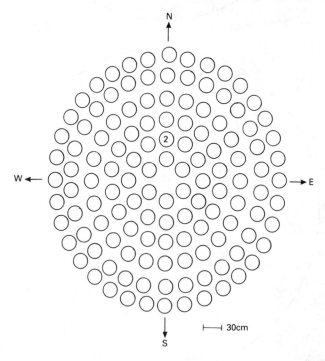

Figure 1.9 *Circular mosaic arrangement of water traps round light trap, in experiments in England (after McGeachie, 1987).*

moths (expressed as a percentage of total moth yield in all traps for a particular evening) in the control trap, and the mean wind speed on evenings when the control trap was downwind of the experimental trap. This was to determine if the two light traps were affected by prevailing wind conditions. Analysis was based on the three most abundant species into the following categories: (i) all moths; (ii) all noctuids; (iii) *Apamea anceps* (noctuid); (iv) *Hepialus lupulinus*; and (v) *Chrysoteuchia culmella* (pyralid). All three named species are grass feeders in the larval form. The noctuid and hepialid named species are large (wing span 25−40 mm) compared with the smaller pyralid, *Chrysoteuchia*, with a wing span of 19−22 mm.

On the calmest evening of trapping, when the mean wind vector was 0.08 m sec^{-1} from 242.08 degrees, moth catches in water traps tended to cluster in the southerly quadrants (Figure 1.10). On evenings with moderate wind speeds of several metres per second, conditions are different. For example, at 3.16 m sec^{-1} from 218 degrees, moth catches in the water traps were clearly biased towards the north-eastern section, i.e. the mean vector was significantly displaced from its original (Figure 1.11). Analysis of data confirmed that wind direction determined the direction of moth displacement from the light trap.

With regard to the total experimental light trap (plus water traps) versus control light trap, there were highly significant differences between some of the categories listed above. For 'all moths' this figure was 2.58 times the control figure (Figure 1.12), but in the case of *C. culmella* this figure was 8.2. From this it was concluded that for that species, catches in the control light trap represent only a small fraction of all the moths attracted to the immediate vicinity of the trap. In the case of the categories 'noctuids', '*A. anceps*' and '*H. lupulinus*' there was no significant difference between experimental total and control. It was concluded that for those categories, most moths appearing in the water traps would have reached the experimental light trap had the water traps not been in operation. In contrast, with *C. culmella*, the water traps sampled some moths which would never have reached the light trap.

(b) Interfering light traps
Using the same standard light source − a 125 W MV bulb − a circular mosaic of water traps was arranged around each trap, as follows:

Distance from lamp in cm	Traps per circle
30	6
60	12
90	18
120	24

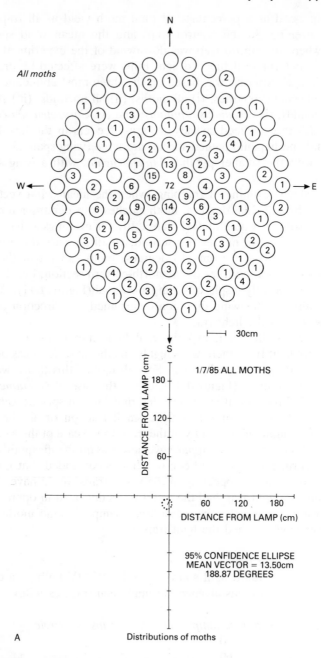

A   Distributions of moths

Figure 1.10 *Distribution of moths around light source on calm evening. A. All moths; B.* Noctuidae; *C.* Apamea anceps; *D.* Chrysoteuchia culmella *(after McGeachie, 1987).*

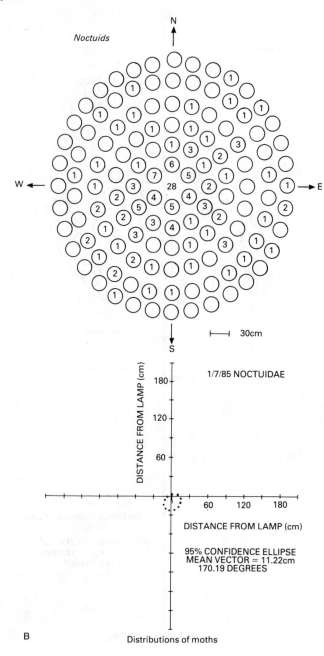

Figure 1.10 *Continued.*

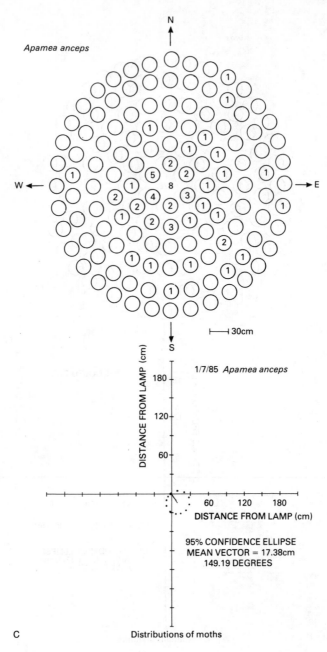

C   Distributions of moths

Figure 1.10   *Continued.*

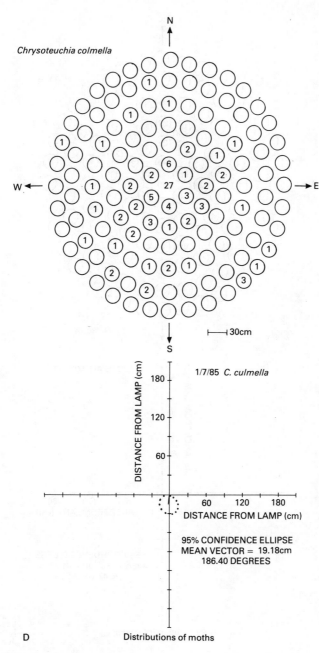

Figure 1.10 *Continued.*

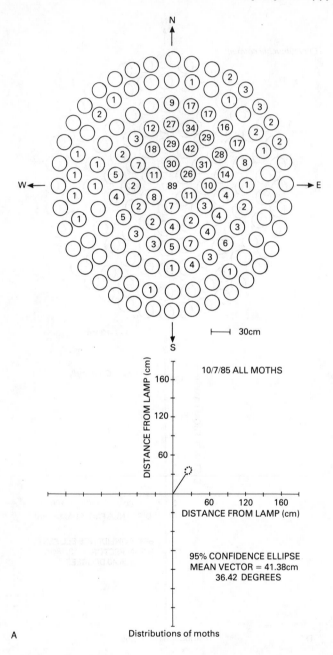

Figure 1.11 *Distribution of moths around light source on night of moderate wind speed (3–16 m s⁻¹). A,B,C,D as in Figure 1.10.*

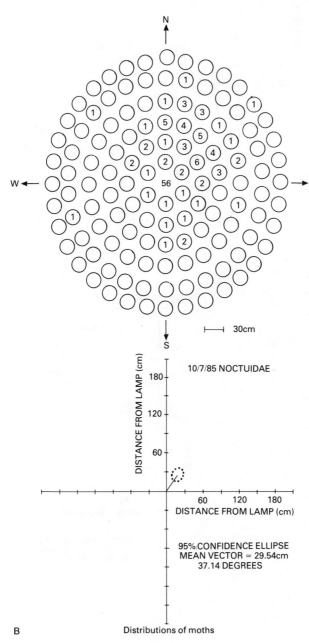

Figure 1.11 *Continued.*

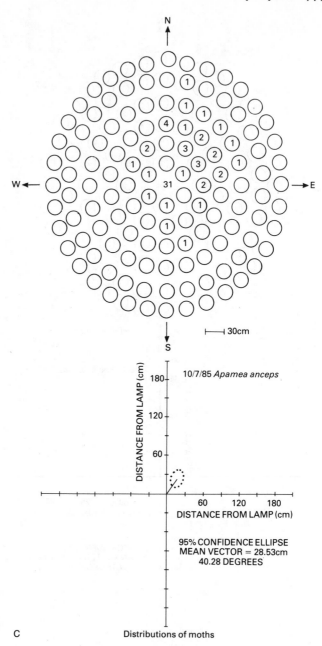

Figure 1.11   *Continued.*

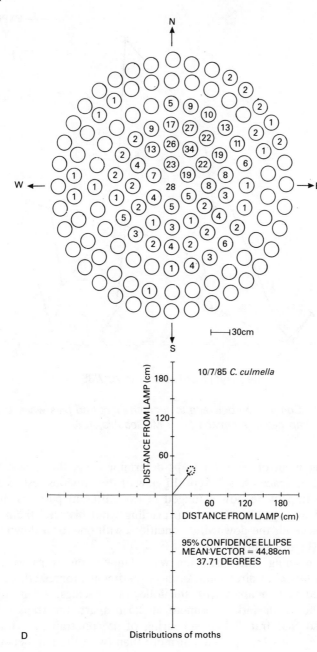

Figure 1.11 *Continued.*

Trap responses of flying insects

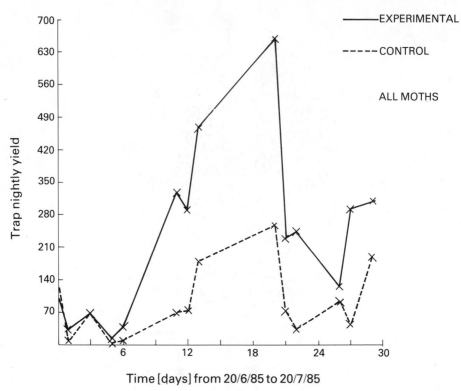

Figure 1.12 *Comparison between total catch (light trap plus water traps) and control light trap catch. All moths (after McGeachie, 1987).*

This arrangement of 60 water traps determined that the closest the two traps could be spaced was 225 cm. In view of the previous results on wind effect, the traps were now arranged in two different patterns: first, the traps were arranged so that the prevailing wind bisected them (Figure 1.13) and second, the downwind condition, with one trap downwind of the other (Figure 1.14).

On the evening when trapping was planned, the traps were set in relation to the wind direction. As the experiment progressed, light traps were placed further apart with the following spacings, 2.5 m, 10 m and 25 m. In the downwind experiment at 2.5 m apart, the traps were first arranged so that trap 2 had a bearing of 45° on trap 1. With trap 1 downwind of trap 2, trap 1 caught more than twice the number of moths than trap 2. This raised the question of whether this downwind displacement of moths was purely a wind effect, or whether it reflected a difference in efficiency of the two trapping systems. This query was answered by

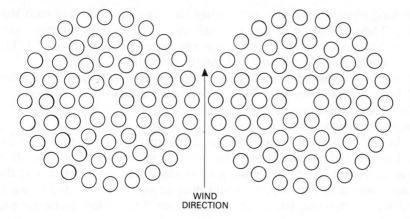

Figure 1.13 *Cross wind arrangement of traps (after McGeachie, 1987).*

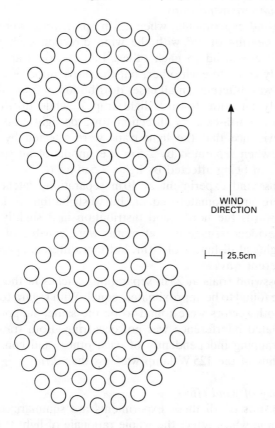

Figure 1.14 *Downwind arrangement of traps (after McGeachie, 1987).*

making observations on a night when the wind direction was such that trap 2 was downwind of trap 1. In this case the trend was reversed, with trap 2 taking nearly three times as many moths (603) as in trap 1 (236), thus confirming the wind-related interference effect.

In the downwind experiment at 10 m apart, with trapping system 2 downwind of system 1, catches were 97 compared with 35. However, as the moth distribution in each system was the same, it appears that 10 m might be approaching the limit of effective trap radius.

In the downwind experiment at 25 m apart, with trapping system 1 downwind of system 2, trap 1 captured 335 moths compared with 109 in trap 2. However, the distribution of moths within each system was the same, indicating that traps were still close enough to be affected by wind-related interference, but far enough apart for light-related interference to disappear.

(c)   Tests with crosswind traps

In the crosswind experiments, when 2.5 m apart, traps were arranged as before at a bearing of 45° with the wind direction such that trapping systems were crosswind to each other. In this case trap catches were approximately equal to each other, but the distribution of catches within each system was different. In all cases distribution in 2 was displaced in a south-westerly direction towards light trap 1, while distribution 1 was displaced in a north-easterly direction towards light trap 2. This was conclusive evidence that the light traps were affected by light related interference when separated by 2.5 m, i.e. moths of all species studied were capable of being affected by lights 2.5 m away.

In the crosswind experiment at 10 m separation, catches in systems 1 and 2 were approximately equal, but distribution in 1 was slightly displaced towards the north, and distribution in 2 slightly towards the south. The evidence suggests that moths were capable of reaching the individual lights at distances of 10 m, but that different species tended to react in different ways.

In the crosswind trials at 25 m separation, catches in the two trapping systems were found to be approximately equal. With regard to distribution, the mean moth vectors were found to be the same, supporting the idea that light-related interference was not occurring. Both trapping systems were now trapping independently of each other, and consequently the effective radius of the 125 W lamp was less than 25 m.

(d)   Summary of wind effect

When the results of all these experiments are summarized, some conclusions emerge which affect the whole rationale of light trapping. From the first series of experiments it was shown that light trap captures are strongly affected by the wind vector. Moths are displaced downwind of a light trap, and — as was shown in the second series — were shown to bias the catch of a second light-trapping system further downwind. The interference is still evident up to a 25 m separation of the two trapping systems. In any comparison of different types of trap, this factor must be taken into consideration, and it is suggested that traps must be separated by distances much greater than 200 m.

Experiments with trapping systems in the crosswind position proved useful for determining effective trap radius, a parameter about which there has been much speculation and controversy. By eliminating the effects of prevailing wind conditions, with both trapping systems experiencing a similar wind vector displacement of moths, differences in moth catch and distribution could be definitely attributed to light-related interference. It appears that in the case of the 125 W MV light trap used in this series, the effective radius exceeds 10 m in most cases, but is less than 25 m. This is a figure much lower than estimates made by other workers on light trap responses of moths. Overestimation of the effective trap radius could be based on the erroneous assumption that moths are attracted to light simply because they can see it. But the present observations show that moths will ignore lamps, despite being able to see them, until at a certain distance from them (10−25 m) they change their flight behaviour (McGeachie, 1987).

1.3.5 Influence of weather factors on light trap performance

The influence of weather factors has been the subject of another recent investigation in England, in many ways complementary to the work just reviewed above (Gaydecki, 1984). In this study, in order to maximize the effect of weather variables, trapping was carried out in an exposed location rather than in a wood or garden. Previous light trapping work has mainly been done in sheltered locations, but such sites not only reduce the impact of winds, but give rise to air turbulence which is difficult to analyse. Also, light traps in wooded areas tend mainly to catch local fauna. In this study the Robinson-type trap was again used, with a 125 W MV lamp, sited 1.4 m above ground. Five species of moth received particular attention, selected not only for their similarity of flight period, but also because they represented a range of sizes which could be related to activity. The smallest moth was *Agriphila tristella*, with a wing span of 23−26 mm, and the largest *Noctua pronuba*, with wing span of 45−55 mm. These five noctuids comprised 62% of the total of over 30 species identified.

The results of observations over a 3 week period, presented in histogram

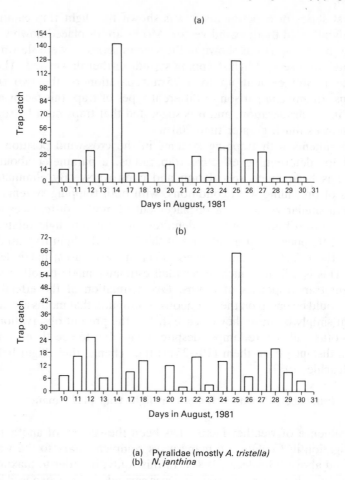

(a) Pyralidae (mostly *A. tristella*)
(b) *N. janthina*

Figure 1.15 *Variations in light trap catch of moths during the course of one month (after Gaydecki, 1984).*

form (Figure 1.15) show how catches changed from night to night. All of these histograms had peaks on the 14th and 25th August, the amplitude bearing an inverse relationship to the size of the insect. The results also showed that for many species the period of highest trap catch was 21.30 to 22.30 hours. These peaks were correlated with mean nightly temperature, and inversely related to nightly mean windspeed (Figure 1.16). These two elements by themselves account for over 90% of the variance associated with the nightly catch. There were also implications that the smaller species are more sensitive to fluctuations in air temperature than larger

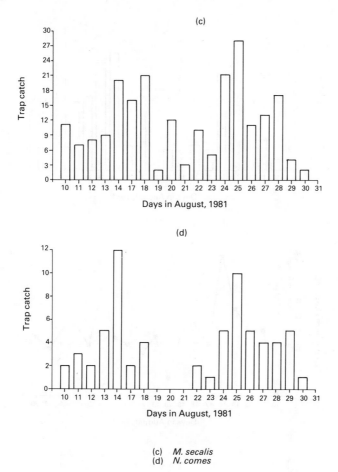

(c) *M. secalis*
(d) *N. comes*

Figure 1.15 *Continued.*

ones, but there was no such obvious relationship in the case of wind speed.

The importance of air temperature in influencing light trap catches had already been stressed in an earlier investigation based on analysis of data obtained during a 10-year light trap survey in western North America (Hardwick, 1972). The mainly noctuid moths concerned — cutworms in particular — were recorded to be less active on cool nights than on warm nights, and that compared with this factor the lunar cycle had no comparable influence on light trap catches.

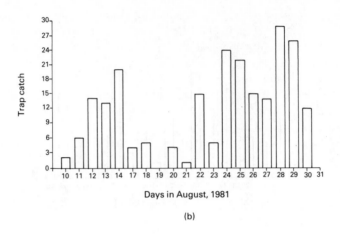

(a)

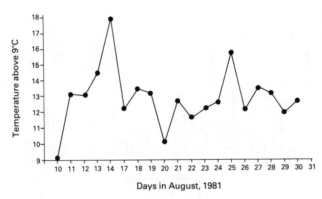

(b)

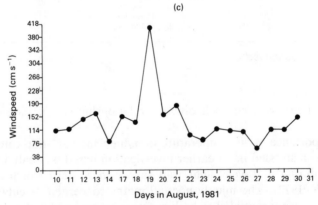

(c)

N. pronuba, windspeed and temperature, August, 1981.

(a)   N. pronuba
(b)   Nightly mean temperature
(c)   Nightly mean windspeed

Figure 1.16 *Correlation between moth peaks, temperature and wind speed (after Gaydecki, 1984).*

In the course of the 3 week observation period, mainly concerned with temperature and wind speed, the full moon was on the night of the 14th August. That particular night witnessed the second largest catch for all species combined, over the 3 week period, as well as the highest number of *Agriphila tristella*, and the second highest number of the noctuid *Mesapamea secalis*. But that night also marked the highest mean temperature and the lowest mean wind speeds recorded.

1.3.6 Moth response to light traps at close quarters

The nature of moth response to light traps at close quarters has intrigued and baffled generations of entomologists. This is well exemplified by two studies separated by 10 years. The later study was able to make full use of technological advances during that period. In the earlier study (Hsiao, 1973) cardboard baffles, painted matt black and coated with a sticky layer of adhesive, were attached to a vertically mounted 15 W Blacklight tube. Moths trapped were identified, enabling a radial density distribution to be determined for each species. The results (Figure 1.17) showed that the number of trapped insects first increased with distance from the lamp, reached a maximum and then decreased. The shape of the distribution and the distance of the maxima appear to vary with species of moth. The cabbage looper, for example, peaked at 45 cm, while the beet armyworm peaked at around 23 cm.

In a later study, in Britain (Gaydecki, 1984) moth flight tracks in relation to light traps were studied by means of TV cameras, provided with an image intensifier in order to achieve video amplification. The most noticeable feature of these tracks was disorientation, often accompanied by rapid changes in speed and changes in pattern, for example helical flight switching to sinusoidal weaving. It was concluded that moth response to light traps is not simply a phototactic one, and in this connection it was noted that moth density fell to a minimum immediately around the trap.

In the second part of that investigation, calibration of light trap catch was attempted by illuminating a known volume of air space with infra-red radiation, undetectable by Lepidoptera. Using the same TV camera fitted with a wide angle lens and an image intensifier, images were displayed and at the same time video-recorded. The area covered by the scan varied from $10.5\,m^2$ to $15.19\,m^2$. The volume sampled was calculated at $25.7\,m^3$. A count was kept of each moth which entered the field of view, its time of appearance and the time spent in the sample area. The results endorsed the previous findings in showing that phototaxis is inadequate to explain moth response to light. Of the total moths flying round the lamp, maximum

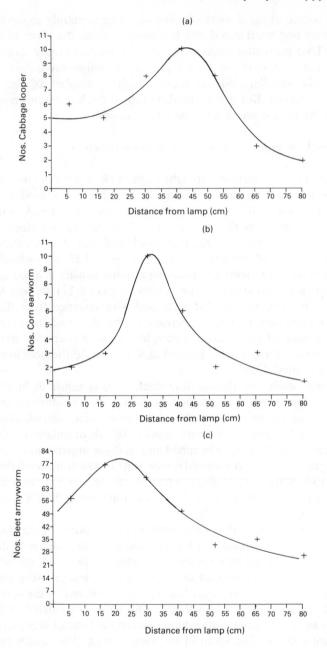

Figure 1.17 *Radial density distribution of three species of moths on baffles round light trap (after Hsiao, 1973).*

density was attained at a distance of about 40 cm from mid-point of the lamp. The study of the moth flight tracks showed the pattern of disorientation is manifested by increased angular velocity with simultaneous positive and negative acceleration, on approaching the light. Minimum speeds took place must often when a moth was flying towards the light, and the converse was often true of maximum speeds. It was also recorded that airspeeds of moths increased with increasing wind speeds as a result of the time spent in the downwind area of the light trap.

1.3.7 Experiments with the New Jersey mosquito light trap

As has been mentioned earlier, the basic design of the New Jersey trap has changed little since its acceptance as standard practice in the USA over 50 years ago (Figure 1.5). Since that time an immense amount of information has accumulated about its performance, and about the range of mosquito species readily captured in such traps, as well as about those species only attracted to a limited extent. Much of that information will be examined later (p. 52). For the moment, the particular aspect of trap performance relevant to this section is the experimental study of mosquito reaction to the original New Jersey trap design, as compared with a slightly modified version in common use. It will be recalled that the New Jersey trap, unlike the designs commonly used for trapping moths, incorporates an electrically driven fan which draws air down from the light bulb area into a container or collector when the trap is in its usual vertical position. However, by inverting that design, the fan now operates above the light source so that insects attracted to the light are sucked up into the collector, and not down. This modification has also been tested out with the smaller, battery-operated, CDC light trap for mosquitoes, which also incorporates a fan. Consequently, any comparative field trial of these two designs has to take into account both updraft and downdraft models (Sexton *et al.*, 1986).

The need for experimenting with an updraught model arose from the fact that in the original downdraught model, beetles and other heavy-bodied insects are drawn into the trap, where they damage the more fragile insects. In order to determine how mosquitoes would react to these two traps, experiments were carried out in the USA with insectary colonies of two species of anopheline mosquito (Wilton and Fay, 1972). Tests were carried out at three different air speeds, 171, 137 and 110 m min^{-1}. The results showed that in the original downdraft model the proportion of exposed mosquitoes captured did not exceed 28%. In the updraught model, the proportion captured ranged from 42% to 72%, depending on the elevation of the traps. Furthermore, in the downdraught model, reduced air flow reduced the catch of both males and females, while in the updraught model, reduction of air flow — within the range of the test — actually increased the catch of females, until the air speed was reduced by half.

An explanation of these findings was provided by experiments in which the flight of mosquitoes dusted with fluorescent powder was observed by UV light. Observations through the window of the test chamber during the operation of a downdraught model showed that individuals which approached the light, but managed to avoid being drawn into the trap, characteristically escaped by flying upwards. That this upward flight would appear to be the normal reaction of mosquitoes encountering an air stream is borne out by observations on the updraught model which showed that the flight path was again upwards and not downwards as might be expected if the reaction was solely against the air flow. In the trap of conventional downdraught type, the air stream has to overcome the lift factor in mosquito flight. In contrast, if the trap produces an upward-moving air stream, the mosquito flight reaction will increase the chances of capture, instead of assisting its escape.

It is worth noting that although both New Jersey light traps and CDC traps have been used for other small night-flying Diptera, such as biting midges (Culicoides) and sandflies (*Phlebotomus*), in the USA Caribbean and Central American region, the former design was found to be ineffective in trapping European species of *Phlebotomus* in the classical studies in southern France (Rioux and Golvan, 1969). This failure was attributed to the fact that the air movement at the fringe of the fan actually repelled flies which had been attracted to the light, before they came within effective range of the fan. The French workers were able to overcome these light trap problems in an ingenious way, namely by installing low-intensity light sources (from a pocket torch) behind semi-transparent sticky paper; traps were set into natural cavities in walls, or strung along like 'garlands'. A further refinement was to standardize light intensity by connecting all pocket torch bulbs to a 6 V accumulator which could maintain constant illumination for many hours of the night, and could also be set to switch on and off automatically.

1.3.8   Light trap responses of night-flying locusts and grasshoppers

A very instructive example of the sometimes highly sensitive relation between insect behaviour and light-trap efficiency is provided by Australian work on locusts and grasshoppers. Light traps of conventional design have long been known to be ineffective in trapping locusts and grass-hoppers, even though these insects are attracted to light from a distance.

This failure has been attributed to strong repulsion to light at close range. Many acridiid species have been observed to be attracted to light, but to alight at some distance away from the point source of light. This led to the design of a trap in which the light was centred above a large, circular, water-filled tank, 10 feet (300 cm) in diameter and 20 in (50 cm) deep. The inside of the tank is coated with white polyurethane paint (Farrow, 1974). The light source is a 150 W lamp with a high output of UV, connected to the mains by an insulated cable. The light bulb is protected by an inverted Pyrex beaker.

The numbers of the Australian plague locust, *Chortoicetes terminifera*, caught in this way ranged from a normal 0−40, to over 15 000 per trap night, these higher figures being recorded on nights of mass flight, such as those associated with disturbed weather (Farrow, 1977, 1988). There is increasing evidence that night flights of solitary locusts and grasshoppers may be as extensive as those of the more conspicuous gregarious locusts (Farrow, 1984).

1.3.9 Light electrocuting traps

Attraction to light has long formed the basis for trapping certain flying insects, particularly house flies. These traps contain one or more fluorescent tubes emitting ultra violet (UV) light or 'black light'. Insects attracted to the light fly into high voltage electrified grids and are killed. These electric traps are marketed mainly for mosquitoes, blow flies and house flies, but particular attention has been given to the responses of house flies to such light sources and traps (Skovmand and Mourier, 1986). In studying the many different factors likely to affect trap response, physiological and environmental elements were tested, both in laboratory studies and in field trials in dimly illuminated animal shelters. The results indicated that in general the catches of very young flies were poor compared with those of older flies, aged more than 2 days. This age difference appeared, however, not to be due so much to differences in physiological condition, but to the fact that younger flies are less active than older ones, and show less exploratory or investigative behaviour than the older flies. In simplified laboratory tests, dealing essentially with light reactions, the importance of these investigative factors is less obvious as the environment of the test chamber is controlled. But in the field, the house fly's well-known attraction to 'new objects' finds full expression, and may override or suppress the attraction of the light trap. This was clearly demonstrated when a white laboratory coat, previously used in a pig shed, was introduced into the controlled environmental chamber, which had a capacity of 21.5 m³. The flies spent much time investigating the coat, and thus avoided death by trapping. The effect was to slow down the capture rate of both sexes of the house fly, *Musca domestica*, in the light trap. The LT_{50}, or time taken to kill 50% of the flies, increased from 13 to 25 minutes in the case of males, and from 27 to 45 minutes in the case of females, all flies being a uniform 7 days old.

In the less controlled conditions of the animal houses used in field tests, another variable factor affecting trap efficiency was disclosed: variations in the light intensity of the trap relative to the background light of the environment, especially during the daylight hours of maximum fly activity.

The cumulative impact of all these variables influencing fly response to electrocuting light traps almost certainly accounts for the many conflicting and inconsistent results of experimental work in the field. Many modifications of such traps, claimed by the designers to increase efficiency, have not proved effective. These traps capture only a negligible proportion of the exposed fly population.

1.4   Moonlight and light trap performance

1.4.1   Introduction

Of all the environmental factors likely to influence the efficiency of light traps, attention has long been given to the effect of moonlight and the lunar cycle. One of the most consistent facts emerging from the long series of observations at Rothamsted, the original centre for studies on light trapping of moths, was that the trap catches were at their lowest around the period of full moon, and at their highest at periods of no moon or new moon (Williams, 1936). With only one or two exceptions, this experience has been confirmed in nearly all light trap investigations carried out since then in many different countries and under a variety of climatic and faunal conditions.

There are two main possible explanations for this. The first is that light trap catches are a true reflection of insect density — or insect activity — and that the low catch at full moon period directly reflects minimal insect activity around that period of the lunar cycle. The other main alternative explanation is that reduced catches at bright periods of moonlight are due to the reduced attraction of the fixed-brightness light trap in competition with the increasingly bright moonlit background. An early resolution of that problem would probably have been achieved if light trap capture data could have been compared with data from an effective alternative sampling method for moths, not involving light as attractant. A comparative series of moth trapping based on some quite different principle would very likely have confirmed whether the declining light trap catches accurately reflected a reduction in moth density, or was an artifact due to reduced trap efficiency at that lunar period. However, the question remained speculative and unresolved for many years mainly because there was no alternative sampling method which could capture numbers of moths comparable with the high catches recorded in light traps (Brown and Taylor, 1971).

Mention has already been made of the experience of light trapping in Africa (Brown *et al.*, 1969) in which over a period of one month, a light trap collected over 5000 African army worm moths (*Spodoptera exempta*) while over that same period suction traps sited 1.5 m and 15 m above ground level captured only three moths. For a long period therefore, research on light trap performance and on the interpretation of light trap capture data, had to depend on studies on light trapping alone, carried out over a wide range of environmental conditions, particularly with regard to phases of the lunar cycle. Extension of light trapping studies to other geographical regions such as the USA, India, Australia and in particular the African tropics also extended knowledge of light trap reactions of moths to a wide range of species, including many important agricultural and forestry pests. In the course of that work, critical studies on all relevant aspects of moonlight and the lunar cycle in relation to light traps provided an essential baseline for unravelling this complex ecological problem.

1.4.2 Moonlight and the lunar cycle

In the course of the regular 28 day lunar cycle from new moon, through full moon, back to new moon again, night-flying insects are exposed to a whole spectrum of variables involving the light factor. First is the progressively increasing intensity of illumination from new moon to the maximum at full moon. Allied to this there is also an increase in the duration of moonlight, from a very brief period at new moon, to illumination throughout most of the night at full moon. Accompanying these changes there is also a sharp increase in the angle of illumination to a maximum of 90° when the full moon is at its zenith.

In northern latitudes, such as those at Rothamsted, where so much of the classic work on light trapping of moths was carried out, at 52°N, there are additional complicating factors due to the wide differences in length of day and night between summer and winter. Much more extensive and variable cloud cover in those latitudes also presents an additional difficulty in interpreting the moonlight factor. The pattern of moonlight is also more complex due to variations in the angular elevation of the moon, and there are also marked seasonal changes in the numbers of insects (Bowden, 1973).

Attention has therefore been drawn to the advantages of studying the lunar cycle in the tropics, where the lengths of day and night are more constant, and where each lunar cycle follows a similar pattern throughout all seasons. Near the equator there is only a small shift in times of moonrise and moonset throughout the year, and the moon rises and sets almost vertically (Bowden, 1973). Long-term studies in Ghana in W. Africa, at Kampala in Uganda and at Mugugu near Nairobi in the Kenyan highlands have enabled accurate figures to be attached to the light factor throughout the complete lunar cycle (Bowden, 1982, 1984).

Because of the more stable conditions at and near the equator it has been possible to compile a table recording the amount of moonlight for each hour of the night, throughout a standard lunar cycle, applicable to any locality between 10°N and 10°S. Thirty-two groupings of moonphase have been defined, within each of which average illumination, and its distribution throughout the night, are almost the same every month and for any site within 10° of the equator. The range of average illumination in these phases is shown in Table 1.2 (Bowden & Church, 1973) showing that the light value, i.e. lux $\times$ 10^4, ranges from a minimum, 9, from the light of the stars only, to a maximum of over 2000 at full moon. When the night is subdivided into four periods according to hours after sunset, the wide differences in illumination between early night and late night are clearly indicated in the phases leading up to full moon. For example, illumination in phase groups 27−29 (last quarter) and 5−7 (new moon) is not much more than in groups 30−34 (no moon), but nearly all the extra light is in hourly periods 7−11 and 1−3 respectively.

Table 1.2 Average moon illumination (lux $\times$ 10^4) at different times of night, by moon phase group.

Phase groups	Hours from sunset			
	1−3	4−6	7−9	10, 11
25−26	9	9	112	166
27−29	9	9	21	50
30−34 (no moon)	9	9	9	9
5−7	45	11	9	9
8−9	179	67	9	9
17 (full moon)	1518	2009	1909	1087

(After Bowden, 1973)

Other light trap investigations have also emphasized that one of the most characteristic features of the light curve of the moon is its steepness, particularly the rapid increase in light intensity towards full moon, and the equally rapid diminution afterwards (Danthanarayana, 1976) (Figure 1.18). The half moon gives only about one-tenth the light of the full moon.

1.4.3 Influence of moonlight and moon phase

From the table of illumination at different moon phases (Table 1.2) an equivalent table can be compiled of radii at which trap illumination equals that of the background, for each phase group and each period of the night (Bowden and Morris, 1975). As the amount of illumination is in lux, radii are expressed in metres. On the basis of the illumination provided by the 125 W mercury vapour (MV) bulb used in light traps in the Uganda/Ghana experiments, i.e. 2900 lm, the maximum radius at which this illumination equals that of the background during the dark period of the new moon has been calculated as 519 m. The minimum radius, which occurs at the lightest period of full moon, is 35 m. The ratio of maximum to minimum is therefore 15:1. These varying effective radii, from 35 to 519 m, really represent circles of influence, determined by the radii. Trap catch is considered to be a function of the frequency with which insects cross the boundary, or the circumference, of this area. Beyond this region of influence, insects are unaffected by the light trap, and they remain so until their movements bring them across the boundary of the region.

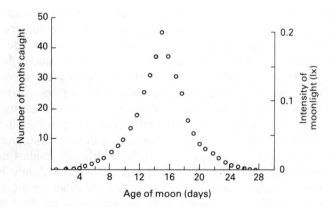

Figure 1.18 *Changes in light intensity of moonlight according to lunar phases (after Danthananarayana, 1976).*

When this concept is applied to whole-night trap data, allowance has to be made for the fact that in the second half of a lunation, i.e. full moon to new moon, illumination is less than in the first half, i.e. new moon to full moon. Trap radii are accordingly greater in the second half. Allowing for this, an index has been calculated summing the individual hourly radii, and expressing totals for each phase group as a proportion of the total at new moon set = 10. This index represents the relative average radius of the region of influence from night to night through a lunation.

The impressive record of light trap data from Ghana and Uganda show that for a wide range of insect species there are wide differences between the proportion caught at no moon and full moon. For the great majority of taxa the general pattern conforms to that established in earlier light trap studies on moths, with catches being 3–4 times higher at new moon than at full moon. But notable exceptions to this general pattern were provided on the one hand by bostrychid beetles in which about three times as many were caught at full moon than at no moon, and on the other hand by a pyralid moth, *Marasmia trapezatus*, which showed exceptionally high catches at no moon, ten times that at full moon. All-night catches showed that some groups are most affected by illumination in the second half of the night, e.g. sphingids, *Heliothis* and *Earias*, while others are more affected in the early part of the night, i.e. less being captured before full moon than after full moon on nights of similar average illumination, such as cantharid beetles and termites.

1.4.4 Environmental factors and light trap catches

In several of the major studies on the reactions of nocturnal insects, particularly moths, to light trapping, observations have embraced a wide range of species whose separation from the trap catch, and whose identification, have all made heavy demands on time. In other studies however, attention has been mainly focused on particular pest moth species, and usually with a specific practical need in mind. This applies, for example, to the extensive studies in East Africa on the African army worm, *Spodoptera exigua*, on the noctuid moths *Heliothis* (now *Helicoverpa*) *armiger* and *H. punctiger*, major pests of cotton in Australia, and on the cotton bollworm, *Heliothis zea*, in the USA. Although studies on those and related species have had a clear practical objective, they have all made important contributions to advances in knowledge on the effect of environmental factors — in addition to moonlight — on light trapping and the interpretation of light trap data.

(a) Experiments on army worm, Spodoptera

In East Africa, as well as in Zimbabwe and South Africa, the occurrence of marked seasonal outbreaks of the army worm has long been recognized (Brown *et al.*, 1969; Douthwaite, 1978). Forecasting such outbreaks is heavily dependent on nightly catches of adult moths in light traps, which have been used regularly from 1963 onwards. The particular method of using the light trap model adopted — the modified Robinson trap with 125 W MV bulb — has already been described (p. 12). The segregation of catches throughout the night has enabled a clear picture to emerge about the trapping rate according to the lunar phases, a rate which, of course, may or may not be associated with changes in population density or activity (Figure 1.19). These results show the weight of insects, including *Spodoptera*, caught at Mugugu, near Nairobi, in 12 hourly periods from sunset to sunrise for four 7 day periods before full moon, and four corresponding 7 day periods after full moon. The results show that in the period before full moon, when the moon rises early and sets long before dawn, more moths are caught in the later part of the night. After full moon, when the moon rises late and is still up at dawn, the reverse is true, with more moths being taken in the early part of the night. Again, the immediately obvious explanation is that the moon competes with the light source from the trap, and renders it less effective (Siddorn and Brown, 1971).

Wind speed, wind direction and rainfall as factors likely to affect light trap catches have received particular attention with *Spodoptera exempta*.

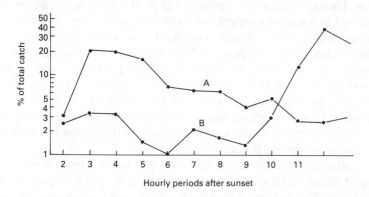

Figure 1.19 *Weight of insects caught by light trap in East Africa over 12 hourly periods from sunset to sunrise for 4–7-day periods before full moon (B) and four corresponding 7-day periods after full moon (A) expressed as percentage of total night catch (after Siddorn and Brown, 1971).*

Observations carried out in dark hours of the night, i.e. between sunset and moonrise, and between moonset and sunrise, showed that catches became less frequent as wind speed increased (Douthwaite, 1978). However, very large catches were most frequent in winds of $1-3\,\text{m sec}^{-1}$ and not in calms. Bearing in mind that in the laboratory the maximum flight speed of *S. exempta* is recorded as $1.8-2.2\,\text{m sec}^{-1}$ it would be expected that catching efficiency of the light trap would decrease beyond that point. However, the situation is complicated by the fact that migrant forms of this species — which orientate downwind — may dominate early in the year, whereas, later, most moths will be of local origin, and tend to orientate upwind. Migrant *S. exempta* are also caught in substantially larger proportions on rainy nights than on dry ones. In contrast, the non-migrant allied species, *Spodoptera triturata*, shows no obvious influence of rainfall on trap catch. It appears that both wind speed and moonlight could be of equal importance in affecting light trap catches (Tucker, 1983).

(b) Experiments on Heliothis *spp. in Australia*
Careful analysis of all these factors likely to affect light trapping has shown that closely related species may exhibit sharply different responses to the same factor. This has been found to apply to two of the major cotton pests of eastern Australia, *Heliothis armiger* and *H. punctiger*. In each of the major cotton-producing areas, light traps are operated as a service to growers, providing information about nightly catches of pest species. Hourly trap catches, using a 125 W MV bulb as attractant, were considered in terms of temperature, humidity, wind speed and moonlight (Morton *et al.*, 1981). Bright moonlight was found greatly to reduce the trap catch of *H. armiger*, by 49%, but had no significant effect on *H. punctiger*. Also, wind speeds in excess of $1.7\,\text{m sec}^{-1}$ had a greater suppressing effect on *H. punctiger* than on *H. armiger*. The results of the hourly and nightly analyses were used to estimate changing seasonal abundance of the two species. But, as was the experience with *Spodoptera* in East Africa, results were affected by variations in the degree of immigration of moths from outside the area.

In another light trap study in Australia, particular attention was paid to the sex ratio of moth species commonly taken in hourly segregated light trap catches. The nocturnal distribution of the most common pest species (Persson, 1976) followed the same general plan, with females most abundant in the first half of the night and males in the last half. However, in two species, *H. armiger* and *Spodoptera litura*, both sexes were most abundant after midnight. If the temperature was high enough, most species also showed a second weaker peak of female flight activity be-

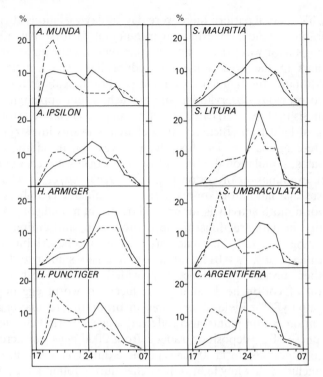

Figure 1.20 *Nocturnal distribution of eight species of male (———) and female (– – – –) moths, overnight in light trap catches (after Persson, 1976).*

fore dawn, a phenomenon which was not only recorded in subtropical Queensland but also in temperate Sweden, and in periods both with and without moonlight (Persson, 1971).

(c) Experiments on cotton bollworm, Heliothis zea, *in the USA*
In the United States, light trap studies on the cotton bollworm moth, *Heliothis zea*, on lines motivated by the pioneer work in Rothamsted, showed that the numbers of moths captured fluctuated according to moon phase, the greater number of moths being captured during the darker new moon phase (Nemec, 1971). The figures obtained showed, for example, that the number of moths trapped per night increased from less than 50 at full moon, to 800 during the new moon period fourteen days later. In the following 15 day period, the catch fell to nearly zero. At peak periods the combined catch of the two light traps used exceeded 1000 per night.

Within this general pattern there were day-to-day variations attributable to temperature, wind and other environmental factors. An additional complicating factor was provided by the fact that insecticidal treatment of cotton plants in the area tended to reduce light trap catches in the vicinity. Although the familiar moonlight pattern appears at first sight to conform to previous observations and as such would appear to confirm the hypothesis that reduced catch at full moon period can be attributed to a lowering of light trap efficiency, new elements in this investigation call for some re-examination. The first element was that in addition to light trap estimates, population densities of the bollworm were also monitored by regular examination of cotton plants for *Heliothis* eggs. The second element was that laboratory observations in bioclimatic chambers showed that bollworm moth activity is suppressed by bright moonlight. Moths are highly active in dark fractions of the photoperiod, and entirely inactive during the bright period. This inactivity continues to be exhibited at light trap intensities as low as 0.01 fc, i.e. a light intensity similar to that at full moon. Other supporting evidence is provided by the fact that artificial illumination of cotton fields at night reduces bollworm egg deposition activity by up to 86%. Field observations incorporating regular egg counts of bollworm on cotton plants also reflected a cyclic periodicity according to moon phase. Egg deposition ranged from 2000 eggs per acre during two full moon periods, to 11 000 eggs per acre during new moon. From all this it is concluded that low light trap catches during full moon periods are a reflection of reduced activity, since reduced egg depositions also occur at full moon periods. In turn, this periodicity in light trap catch and in egg depositions, both correlated with moon phase, is a reflection of bollworm generation cycles.

1.5   Use of additional sampling methods to interpret light trap data

All the work on the interpretation of light trap data reviewed so far has had to be carried out in the face of the major handicap that the light trap has been the sole capture or sampling technique used. From light trap data alone, many speculative conclusions have been reached regarding the nocturnal density, activity and flight patterns of night-flying insects, moths in particular. Many of the tentative conclusions could have been tested, or discarded as the case may be, by applying additional alternative trapping or sampling methods, especially those not involving light as an attractant. This need is strikingly evident in the numerous studies on the effect of the lunar cycle on light trapping, all of which — with only an odd dissident — have demonstrated that high light trap catches at the period of no moon or new moon invariably decrease to the lowest level at the full moon period. The question of whether this represents a real fall in moth density or activity at full moon, or whether it is an artifact attributable to reduced trap efficiency in competition with the full moon, has not yet been clarified. Numerous attempts to explain these phenomena on the basis of light trap data alone have been almost inevitably doomed to failure. In many cases there is really no justification to conclude anything more than that under certain conditions light traps are successful in capturing a larger number and a greater variety of moths than any other known method.

1.5.1 Light and suction traps: *Chrysoperla*

In the case of moths it is only comparatively recently that additional sampling techniques have helped to establish a sounder concept of behaviour patterns and flight paths, as well as changes in population density, all of prime importance in the elucidation of light trap data. But before dealing with these advances later in this section, attention has to be drawn to other groups of insects, readily taken in light traps, which have been more fortunately placed with regard to alternative capture methods. The first of these is the neuropteran (stonefly) *Chrysoperla carnea*, one of the insect species regularly taken in both light traps and suction traps at Rothamsted (Bowden, 1981). A comparison was made between one of the four Rothamsted light traps and a suction trap, located 100 m distant. Geometric mean catches are shown in Figure 1.21, covering eight years. Similarities and differences are best illustrated by representing the catch for each phase group of the lunar cycle as a ratio of the catch at new moon (Figure 1.22). The curves are quite similar — with the light-trap catch somewhat larger — until phase group 11, i.e. just after first quarter. From then on they diverge significantly, the light trap catch declining to a minimum at phase group 17, i.e. full moon, then rising quite rapidly to parity with the catch at new moon, two to three nights after full moon. Just before the last quarter (phase group 25) the catch which has been steadily rising, declines quickly to about half the new moon catch.

The suction trap catch rises to a very sharp peak — almost four times the new moon catch — at phase 13, i.e four nights before full moon. It then declines rapidly to a minimum at full moon; but about three nights after full moon it rises to another peak, and then subsides to a level about half that of new moon. In comparing these two sets of data, it is assumed that the suction trap catch, unlike that of the light trap, is completely

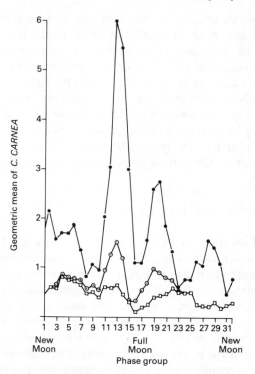

Figure 1.21 *Comparisons between catches of* Chrysoperla carnea *at light trap catches, and by suction trap, over lunar period (after Bowden, 1981).* • *Suction trap;* □ *light trap catch;* ⊙ *light trap catch adjusted to allow for variations of illumination in first 3 hours after evening civil twilight.*

unaffected by background illumination. It follows therefore that fluctuations in the suction trap catch associated with different moon phases cannot be attributed to changes in trap efficiency, but represent the real activity pattern of the insect. As such, the figures do give some degree of support to those who have contended in the past that the fall in light trap catch at full moon is due, in whole or in part, to an inhibiting effect of moonlight on flight, and not attributable entirely to reduced light trap efficiency at that period.

1.5.2 Interpretation of New Jersey light trap data

Comparative studies in Florida and Panama
The New Jersey, mains-operated and fan-assisted light trap has been widely used in American countries for monitoring populations of mos-

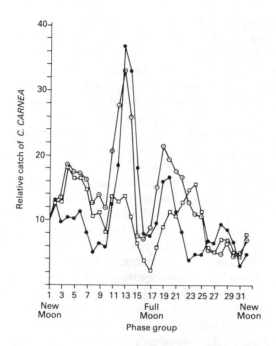

Figure 1.22 *Comparisons in Figure 1.21 for each phase group of the lunar cycle, as ratio of catch to new moon figures (after Bowden, 1981).* • *Suction trap.* □ *light trap;* ⊙ *adjusted light trap.*

quitoes and other small nocturnal biting flies, such as midges (Culicoides) and sandflies (Phlebotomus). In contrast to moth studies, there are several alternative methods for capturing, trapping and sampling these biting insects, in addition to light traps. Consequently, it is usually possible to provide two or three parameters with which light trap data can be compared. It will be sufficient for the moment to quote two good examples, referring to mosquitoes and to phlebotomine biting flies respectively.

In the work in Florida in which several different sampling techniques were compared, the contrast between the New Jersey light trap and suction traps is particularly instructive. Different groups of mosquitoes involved in these trials were arbitrarily classified according to the ratio

$$\frac{\text{New Jersey trap catch}}{\text{Suction trap catch}}$$ during the entire night (Bidlingmayer, 1967).

The first group, with ratios ranging from 8:1 to 1:1 were highly attracted to light, and are exemplified by *Anopheles crucians*. The second group, with ratios ranging from 0.7:1 to 0.4:1, were only moderately attracted to light, the suction trap being more effective. Examples of this group are the brackish-water pest mosquito *Aedes sollicitans*, and the swamp and flood-water breeder *Anopheles quadrimaculatus* (females), the main vector of malaria, which was formerly prevalent in the southern United States. The third group were either not attracted to light traps, or were actually repelled, giving an NJ:suction trap ratio of 0.2:1 down to 0.1:1; the latter figure indicates that on average the light trap catch was only one-tenth of the suction trap catch. Examples of this group were the pest mosquito *Mansonia perturbans* and *Anopheles quadrimaculatus* males.

A similar type of extensive comparison involving light traps was carried out on phlebotomine biting flies in Panama (Chaniotis *et al.*, 1971). In this case the light trap was the small portable battery-operated model, the CDC miniature light trap, well suited for use in jungle conditions in the absence of mains electricity supply. During the 59 week study period a total of over 60 000 sandflies, representing 37 species, were collected in a comparison between light traps operating at two levels, and synchronized night-time collections by aspirator. The results are shown in Table 1.3. From this it can be seen that some species (1,2 and 4) were attracted almost exclusively to light, and among these was the main man-biting species, and suspect vector of virus, *Phlebotomus panamensis*. In another group (3,5,6 and 8) resting captures (made by aspirator) greatly exceeded light trap captures, and light traps clearly had only a low degree of attraction. It will also be noted that in three of the four species taken abundantly in light traps, traps at ground level took higher catches than at canopy level.

Table 1.3 Number of different species of sandflies taken in Panama by three concurrent catching/sampling techniques (Chaniotis *et al.*, 1971).

| | Light trap | | |
Species	Ground level	Canopy	Aspirator
1. *Lutzomyia aclydifera*	2675	513	3
2. *L. carpenteri*	2303	415	2
3. *L. ovalessi*	71	23	551
4. *L. panamensis*	5152	1615	43
5. *L. roratensis*	7	4	632
6. *L. shannoni*	128	44	9799
7. *L. trapidoi*	2092	3128	2026
8. *L. trinidadensis*	233	85	12687

(b)   Studies in Florida involving the lunar cycle

The examples above have contrasted numbers of insects captured by light trapping and alternative methods in the same locality. One measure of light trap efficiency has been highlighted, insofar as different species of mosquito and sandfly are concerned. The figures reveal that when insect response is low, reliance on light trap data alone may greatly underestimate the abundance of any particular species. Further application of such supplementary sampling methods, which do not involve light or other attractant, has enabled a more penetrating analysis to be made of the complex relationship between lunar phase and light trap performance. This is mainly the outcome of a long-term mosquito research project in Florida, which was initiated in the 1940s and has continued until very recently.

It was very early experienced in the Florida programme that mosquito catches were lower at periods of full moon than they were at quarter moon or no moon, in much the same way as was experienced in the Rothamsted light trap work. This finding was interpreted at that time as being due to the fact that at different phases of the moon, the constant light emitted by the New Jersey trap was competing with a range of background illumination, which reached maximum intensity and duration at full moon. Later, in a more detailed study involving seven genera of mosquito, carried out according to the phases of the moon, three of these species were sufficiently abundant to confirm and quantify this conclusion (Figure 1.23) (Provost, 1959). In extended trials in another locality in Florida, the six species of mosquito involved in the light trap trials were being continuously produced from extensive breeding grounds throughout the trial period. Consequently, any wide differences in light trap catch could not be attributed simply to any marked periodicity of output, increasing or decreasing the adult population available for capture. The results (Figure 1.24) amplify the previous findings, and in addition provide new information about sex differences in reaction to light traps.

In a fourth series of trials, in still another locality, an additional sampling method was introduced in the form of the truck-mounted funnel trap (Figure 1.25) designed to study populations of the swamp breeding mosquito *Aedes taeniorhynchus* which were present at high density. These truck traps, described in more detail elsewhere (p. 171) made their run on an 18 mile circuit on which six light traps were in operation. The truck covered the circuit three times a night, starting after dark, at one hour before midnight, and at pre-dawn at 4 a.m. The results (Figure 1.26) show that around and subsequent to the full moon there is a wide divergence between the two trapping methods. The full moon period is marked by a sharp fall in light trap catch on the one hand, and by an

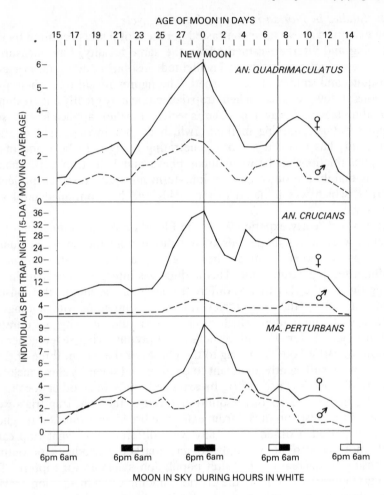

Figure 1.23 *Light trap catches of three species of Florida mosquito according to moon phase (after Provost, 1959).*

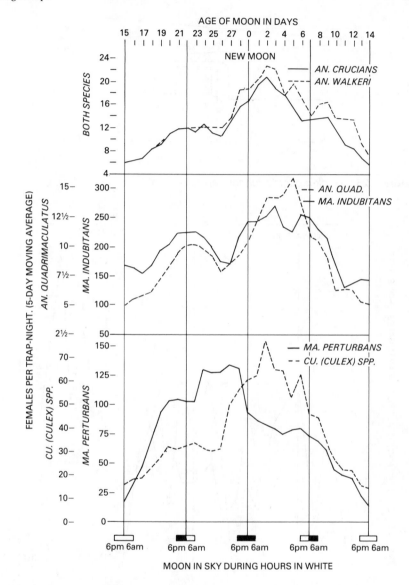

Figure 1.24 *Catches of six species of mosquito — continuously produced from breeding swamps — at light traps according to lunar phase (after Provost, 1959).*

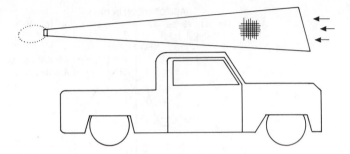

Figure 1.25 *Truck-mounted funnel trap (after Bidlingmayer, 1974).*

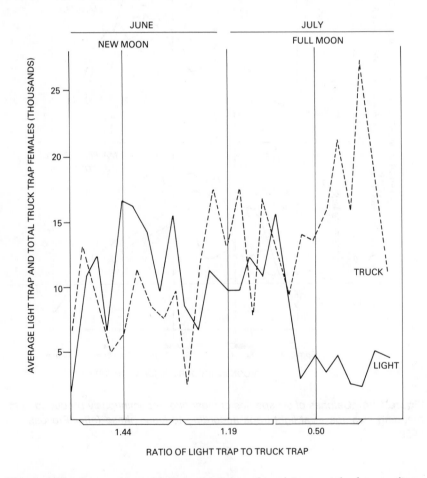

Figure 1.26 *Comparison of light trap catch and truck trap catch of mosquitoes throughout the lunar cycle (after Provost, 1959).*

equally marked increase in the truck trap catch on the other, which indicated a peak of mosquito density at that period.

In a further extended application of the truck mounted funnel trap, regular runs of 15–20 min were made over a short course of approximately 3 miles, from sunset through to sunrise. It was found convenient to divide the night into eight periods as follows:

Evening from sunset	Night time. Dark hours	Morning until sunrise
1	2 3 4 5 6 7	8

Each of the six dark(night) periods was roughly the same duration as the twilight and dawn crepuscular periods, i.e. 75–80 min. In accordance with the lunar light variations it was found convenient, in recording data, to combine periods 2, 3 and 4, and 5, 6 and 7 to form two groups characterized by the presence or absence of moonlight in the quarter moon period. Tabulated in this way, the results show first of all that the greatest flight activity occurred during the illuminated part of the night (Table 1.4) (Bidlingmayer, 1964). Within this quarter moon phase, further analysis of data taking into account that the amount of moonlight will vary with the altitude of the moon above the horizon, showed that both first quarter and last quarter collections in the moonlit half of the night were largest when the moon was near zenith, and lowest just before moonset, or immediately after moonrise.

Comparison between the two extreme conditions of illumination, namely full moon illumination throughout the night and complete darkness (new moon period) through the night, revealed striking differences. On the night of the new moon the number of female *Aedes taeniorhynchus* collected by truck trap remained at a uniform level throughout the six periods of the night, but at a much lower level than either the previous twilight period, 1, or the following twilight period, 8 at dawn. In the case of full moon illumination throughout the night, a uniform level of activity — as judged by truck trap collections — was also maintained through the night, but at a much higher level. This level was between 6 to 8 times higher than in the dark period of new moon (Table 1.4), and at some points approached the peak flight activity recorded at twilight.

When these findings are applied to the earlier problem of the relative efficiency of light traps at different lunar phases, it will be seen that the established fall in light trap performance at full moon is even greater than previously estimated due to the fact that this full moon period is actually one at which there is a maximum number of mosquitoes in flight. With regard to male mosquitoes, the indications from the small samples available

Table 1.4 Mean number of female *Aedes taeniorhynchus* per truck trap collection at different periods of the night (After Bidlingmayer, 1964)

Period	Evening (From sunset) 1	Night 2	3	4	5	6	7	Morning (Until sunrise) 8
		First quarter moon						
		Moonlight			Dark			
	66.6	34.5	29.9	19.0	8.6	6.1	6.6	37.9
		Last quarter moon						
		Dark			Moonlight			
	54.0	7.1	6.8	8.1	13.5	20.9	32.9	52.7
		Full moon						
		Moonlight			Moonlight			
	56.6	32.9	42.7	39.7	48.0	45.8	39.9	32.1
		New moon						
		Dark			Dark			
	38.8	5.3	6.3	5.2	4.3	5.0	4.5	31.4

are that their activity is not increased by moonlight to the same extent as with females, and that other factors operate to concentrate male activity to twilight, even when the light intensity at night is favourable. It is also interesting to note that the general flight pattern of female salt-marsh mosquitoes appears to be unaffected by cloud cover, this fact being no doubt associated with the recording that, on nights with an obscured moon, the light still remains brighter than on moonless nights.

1.5.3 Australian work on the interpretation of lunar periodicity

In the continuing studies on moonlight in relation to flight periodicity of moths, the limitations on research workers by having to rely on a single sampling method, namely the light trap, have become all too evident. The great advantage of using a range of sampling or capture methods for night-flying insects has been firmly established in mosquito research for many years, and is amply illustrated by the work in Florida just reviewed.

But it was not until a comparatively late stage that corresponding progress was made in the long-standing problem of moth flight, namely by fully exploiting the suction trap as a supplementary technique (Bowden and Gibbs, 1973).

The first subject of this critical approach was the light brown apple moth (*Epiphyas postvittana*), a serious pest of many important fruit crops in Australia and New Zealand (Danthanarayana, 1976). Using a 23 cm (9 in) suction trap with hourly segregation mechanism, this species was shown to have a dual periodicity, with a peak of flight activity 2−3 hours after sunset, i.e. around 21.00 hours (Figure 1.27). The use of the suction trap also revealed an additional feature that could not have been detected by light trap, i.e. the existence of a second, smaller daytime peak of activity 3−4 hours after sunrise. These observations were then extended to the entire lunar cycle; the lunar month with its constant duration of 29.53 days was divided into 12 intervals of equal length. When the number of moths captured at each moon age was plotted (Figure 1.28) a trimodal curve was revealed, with peak catches shortly after new moon;

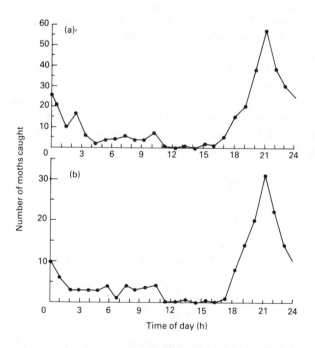

Figure 1.27 *Diel periodicity of flight activity of* Epiphyas postvittana *(males (a) and females (b)) (after Danthanarayana, 1976).*

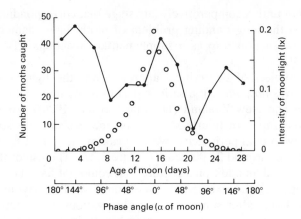

Figure 1.28 *Fluctuations in numbers of* E. postvittana *caught in suction trap at night in relation to moon phase* ● *Number of moths caught;* ○ *Intensity of moonlight (after Danthanarayana, 1976).*

immediately after the exact date of full moon (i.e. day 14.77), and shortly before new moon. The same pattern was shown for both sexes.

Similar trapping methods were extended in Australia to the cabbage moth, *Plutella xylostella*, an international pest of cruciferous plants (Goodwin and Danthanarayana, 1984). Both sexes of this species were also found to show a clear lunar periodicity, with peak flight occurring at full moon, i.e. 15 day old moon, and two minor responses 2−4 days before and after new moon, i.e. days 1 and 29 (Figure 1.29). Both of these species therefore show a distinct trimodal activity influenced by lunar periodicity.

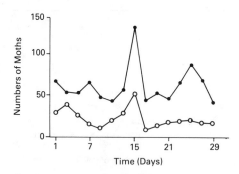

Figure 1.29 *Lunar periodicity of flight activity − males (○−○) and females (●−●) − of* P. xylostella *in suction traps (after Goodwin and Danthanarayana, 1984).*

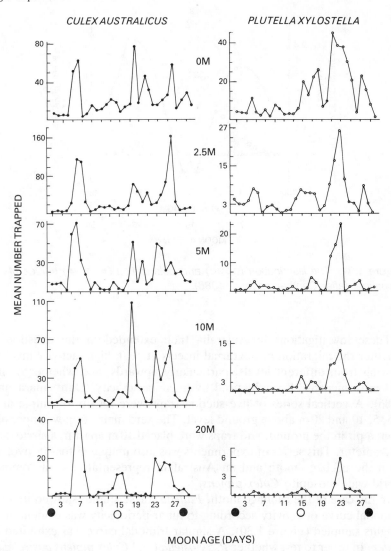

Figure 1.30 *Mean numbers of the mosquito* Culex pipiens australicus, *females, and the moth* Plutella xylostella *(males and females) caught in suction traps per lunar month at each trapping height, based on data from nine lunar cycles (after Danthanarayana, 1986).*

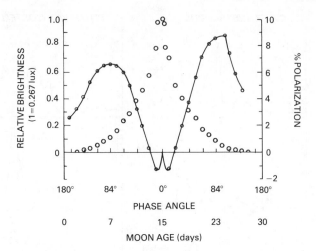

Figure 1.31 *The illumination (○○○○) and polarization (●─●─●) curves of moonlight (after Danthanarayana, 1986).*

These investigations in Australia then extended to the question of whether the migration of nocturnal insects at peak flight activity involves moving from different levels, particularly upwards, and whether or not this movement is also influenced by lunar periodicity (Danthanarayana, 1986). A vertical series of five suction traps were set up to sample at 0, 2.5, 5, 10 and 20 m above ground level. The zero metre trap was operated from a pit in the ground, and traps were placed 10 m apart in a circle 20 m in diameter. This series of experiments was also unique in that it involved both the cabbage moth and an Australian representative of a common world-wide mosquito *Culex pipiens*.

In the case of the cabbage moth, *Plutella*, these studies showed that the trimodal curve of activity according to lunar periodicity was evident at all heights sampled (Figure 1.30). A similar trimodal curve was exhibited by *Culex*. In order to test whether *P. xylostella* (and *Culex pipiens australicus*) enter the upper levels of the air for wind-assisted migration, in relation to the lunar phases described above, the hourly vertical─density profiles for the three periods of peak activity were determined, as well as for the two periods of minimal flight activity. The numbers of insects trapped were converted into aerial densities by allowing for the amount of air sampled by the traps, and their efficiency at various wind speeds. The traps from 2.5 m to 20 m were used, and data was extrapolated to 80 m, since 70 m is regarded as the critical height above which there is usually sufficient wind for migration. The results showed that during the lunar cycle, upward movement of *Plutella* and *Culex* occurred during the three peak periods of flight, i.e. pre- and post-new moon, and full moon, rather than at new moon or during the other trough period.

There is evidence that the trimodal curve, according to lunar phase, established for these two moth species as well as for the mosquito *Culex pipiens*, may be widespread among other nocturnal insects. But if sampling is limited to light traps, this peak around full moon is obscured, either on account of the competition between the light from the moon and from the light trap at that time, or because of the reduced trap efficiency, or to both in conjunction. It is also worth noting that this trimodal pattern could not be explained in terms of environmental variables such as temperature, humidity, rainfall and wind speed, but a correlation was found between the numbers trapped and the amount of polarization of moonlight (Figure 1.31).

Chapter 2

Suction Traps

2.1 Introduction

The extraction and retention of flying insects from a volume of air is the basis of a wide range of sampling and trapping techniques for winged insects. From this common objective, capture methods have developed along two distinct lines. In one direction the capture net is moved through the air, as in the manually held collecting net or 'butterfly' net, or the same objective is achieved when the movement of the net is power-operated, as in the truck or vehicle-mounted trap. In the other direction, the air in which insects are in flight is drawn through a fixed trap by means of a suction fan, as in the conventional suction trap. All these applications of the same basic principle have played an important part in different fields of applied entomology, according to different requirements and different objectives.

The development and application of the suction trap itself has been particularly associated with one major research centre, the Rothamsted Experimental Station, Harpenden, UK for over 30 years. The experience gained over that period has provided a stimulus and a model, not only leading to present sophisticated designs perfected at that centre, but also in the application of suction trap techniques in many other fields, including insects of medical and veterinary importance.

Intensified interest in the development of a non-attractant method of sampling or monitoring aerial populations of flying insects arose in the 1940s with the need for an improved method for the quantitative sampling of aphids and other small airborne insects of agricultural importance. Until then the trapping of aphids had relied mainly on sticky traps and aerial stationary 'tow-nets' (Broadbent, 1948; Broadbent *et al.*, 1948). As there was no way of establishing a relation between the number of aphids caught in the sticky trap and the number in the air, it was imperative to devise a method of determining numbers per unit volume of air. A knowledge of the density of insects in the air, and of the insect populations involved, was essential towards a better understanding of aphid biology in relation to transmission of plant virus diseases. It was also becoming important to devise effective standard methods for detecting aerial immigration of aphids into the UK sufficiently sensitive to detect the low density levels which were precursors to outbreaks of aphid infestation.

The first such traps began operations at Rothamsted in 1964, and have continued ever since (Johnson & Taylor, 1955; Taylor *et al.*, 1981a). Subsequently a network of 23 traps was established throughout Britain as part of the Rothamsted Insect Survey to provide data for both an aphid warning system and on a wide range of ecological studies on dynamics of insect populations. (Taylor *et al.*, 1981a,b Woiwood and Dancy, 1986, Woiwood *et al.*, 1984, 1985, 1986) Similar 12 in suction traps have been adopted in many other countries of western Europe, and an essentially similar model of suction trap network has more recently been established in the western USA (Allison and Pike, 1988).

2.2 The Rothamsted insect survey 12 in suction trap

In the course of its evolution, the design of the suction trap used at Rothamsted has undergone successive modifications and improvements. These different models have all provided essential information about the performance of such a capture technique under different environmental conditions, and about the various factors which determine trap efficiency. However, in view of the fact that a standard design has now been settled, one agreeable to both British and French observers, there would be no real advantage in devoting too much time and space to the design and construction of earlier models, except insofar as these models were the ones on which so many earlier experiments were carried out, experiments which have provided so much of the basic knowledge of this technique.

The present standard trap (Macauley *et al.*, 1988) (Figure 2.1a,b) has two major components: (a) a 9.2 m plastic pipe on top of (b) a 3 m box containing an electric fan and the necessary filter and storage devices to collect the insect sample. The trap inlet is 12.2 m above ground — this being the equivalent of the original round figure of 40 ft — and the internal pipe diameter is 244 mm, with an inlet flare at 30° to the long axis of the pipe. The inlet air speed greatly exceeds that of insect flight, to give a sample volume of 45 m³ min⁻¹. Both parameters for height and air-inlet speed were arrived at on the basis of previous experience (Johnson, 1957; Taylor, 1974) which had demonstrated that insect density tends to decrease with height, and that a trap sampling 40–50 m³ min⁻¹ at a height

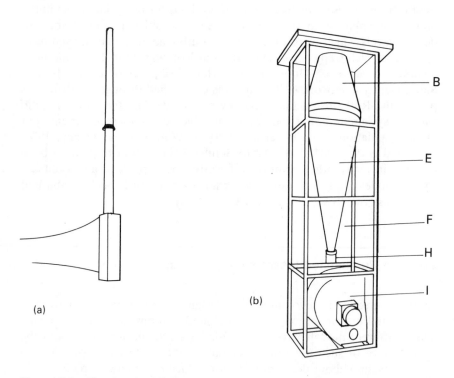

(a)   (b)

Figure 2.1   *The standard Rothamsted 12 m suction trap Macaulay et al., 1988).*
(a) Complete trap with extension to 12 m (b) Box construction of
partly complete 12 m suction trap.
(I. Electric fan. H. Collection bottle for insect samples. E & F. Upper
and lower sections of net. B. Expansion chamber.)

of 40 ft (12.2 m) would be high enough to exclude most insects flying locally, but low enough to sample the densest layer of migrant insects, i.e. those which are of particular concern in the monitoring surveillance. It is important to note that this standard modern trap has a much higher air intake than earlier models, many of which incorporated 9 in Vent-Axia suction fans which sampled approximately 19 000 cubic feet per hour, i.e. 538 m^3, or 9 m^3 min^{-1}.

The standard model is also provided with an expansion chamber to reduce insect air speed, and a sampling net designed in such a way that the air speed through the net is reduced to approximately 1.2 m sec^{-1}. During operation, the air passing down the pipe reaches a velocity in excess of 50 km h^{-1}. The conical expansion chamber is incorporated to slow this air speed in order to avoid insects impacting on the stainless steel net at too great a force. At the bottom of the net a polythene tube leads to a wide-mouthed bottle in which insect samples are collected in a mixture of 68% methanol, 30% water and 2% glycerol. The electric fan is powered by mains 230–250 V, and the particular British design is 'Impellair'. A circular frame above the opening of the pipe prevents access of birds and their droppings. The trap is erected by means of a trailer-mounted hydraulic hoist. Finally, the collecting system incorporates an automatic bottle changing device.

2.3 The Rothamsted suction trap: trap performance and insect reaction

As suction traps at Rothamsted were originally designed to replace sticky traps, earlier experiments comparing these two techniques are instructive (Johnson, 1950), especially in view of the fact that 'sticky' traps are still in use for monitoring certain small insect pests in other parts of the world (Chapter 5). The sticky trap used in the present comparison was 12 in long and 5 in in diameter; it was painted white, and round it a sheet of cellophane with grease banding was fixed. In addition to these two trapping devices, a third was included in this experiment, a stationary tow net. These tests showed that at wind speeds below 3–5 mph, the suction trap captured much higher numbers of small insects — chironomids and aphids — than the sticky trap. Above wind speeds of 7 mph the tow net was more effective as it now sampled a much larger volume of air than the other two methods. In contrast, the tow net was relatively inefficient at low air speeds, even though at air speeds of 2 mph it still sampled a greater volume of air. This was attributed to escape of insects after capture, especially when the mouth of the net faced the sun. In further comparisons of suction trap and sticky trap alone, at a height of $4\frac{1}{2}$ ft, aphid catches were again higher in the suction traps at air speeds below 3 mph, but at this air speed large insects managed to avoid the suction trap, and were more readily taken in the sticky trap.

Early work on suction traps established principles which are still relevant to the evaluation of suction traps in general, particularly if the purpose of the trapping is to measure aerial densities of insects (Taylor, 1962, 1986) It was stressed, for example, that suction traps should be neither attractive nor repellent visually. Visual attraction to a trap would artificially increase the aerial density of insects near the air intake. In order to check this point, the trap was kept as far as possible from the collecting inlet by means of a long extension of semi-transparent plastic tubing, and the results compared with a normal trap. The indications were that, as far as aphids and psocids were concerned, there was no indication of attractiveness.

The possible effect of wind speed on performance was tested by comparing three traps at a wide range of wind speeds. The catch from each trap, on each occasion, was divided by the total quantity of air taken in by the trap at the mean wind speed of the run, and multiplied by one million. This gives the estimated density of insects per 10^6 cubic feet of air. The conclusion from these trials were that, while wind speed does affect the number of insects in flight, wind speed itself does not affect the efficiency of the suction traps. In these tests, any possible avoiding reaction of insects approaching the trap inlet was minimized by extending the trap inlet forward into the wind by means of an extension of transparent material.

2.4 Suction traps for mosquitoes

Suction traps, based essentially on the model developed at Rothamsted, i.e. the Johnson–Taylor trap, have been widely used in mosquito studies, and have been further modified to deal with those special requirements. In many cases the suction traps have been used on their own, while in others they have been combined with an attractant appropriate for biting insects seeking a blood meal, in the form of a bait animal or of CO_2 (dry ice).

Three separate projects within the last 20 years have been particularly illuminating. These are the long-term mosquito research project in Florida, USA and parallel investigations in the Gambia, West Africa, and the studies on woodland mosquitoes in England. Each of these projects studied suction trap performance from different angles, and from quite independent viewpoints.

2.4.1 Mosquito studies in Florida

One of the main themes of this long-standing and continuing study on mosquito behaviour has been the critical assessment of each trapping and sampling technique used in these ecological studies. Particular emphasis has been placed on 'attractant' trapping methods on the one hand, such as light traps, live bait traps, CO_2 traps, etc., and 'non-attractant' techniques on the other, such as suction traps and truck or vehicle-mounted traps, all suitable for insects in flight. In addition, techniques for the mechanical collection of outdoor resting populations of mosquitoes based on the vacuum principle could also be classed as 'non-attractant' (Bidlingmayer, 1967, 1971, 1985)

(a) Experimental studies on trap sites and surrounds
The suction trap used in this work took the form of an L-shaped box panelled trap, with a rectangular inlet 79 × 79 cm on the upper surface, 1.5 m above ground level (Figure 2.5). The fan moves approximately 100 cubic feet of air per minute. In the course of work on these traps it soon emerged that, although they were identical in construction and operation, there were appreciable differences in the size and composition of the mosquito catches according to location. This 'trap site' phenomenon is well known from many other mosquito investigations, the information usually being put to immediate practical use by siting traps in the most productive situation. In few cases had any deeper analysis of this site preference been attempted.

The observations in Florida indicated that mosquito responses were influenced by environmental factors at the different sites, particularly with regard to shrubs, trees and clearings. A series of experiments was therefore designed to examine this, with the L-shaped suction trap being used as standard non-attractant sampler (Bidlingmayer, 1975).

In the first experiment, traps were arranged simply with the object of determining the level of mosquito activity in the different habitats inside, adjacent to and well outside a wooded swamp area. It had previously been shown that while the period of full moon was marked by greater mosquito flight activity than at new moon — and for that reason might have seemed an ideal period for comparative tests — traps in the open were more strongly affected at that period than those beneath the forest

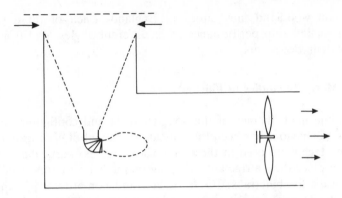

Figure 2.2 *Box-shaped suction trap used in Florida mosquito studies (Bidlingmayer, 1967, 1971).*

canopy. Accordingly, for strictly comparable purposes, traps were operated for five nights only, centred on the new moon.

The next experiment, designed to test the effect of vegetation at the trap site, used shrubs about 8—11 ft tall, planted in steel tubs so that they could be moved from site to site. Four traps were operated: first, without shrubs, in order to test any bias associated with position of trap; secondly, with shrubs arranged in a close circle about 12 ft in diameter around the trap intake; thirdly, with the shrubs placed in a 30 ft row perpendicular to the edge of the swamp, the trap resting in the centre of the row as in a gap in a hedge.

The results of the initial test — without shrubs — confirmed that mosquitoes could arbitrarily be divided into three classes: those which occurred in the greatest number in the open field, with numbers diminishing towards the interior of the swamp; those which occurred in the lowest number in the field, and in increasing numbers towards the interior of the swamp; and those with no well-defined habitat preference.

In nearly all cases, the traps encircled with shrubs took smaller numbers of all species. In contrast, all species were taken in proportionately greater numbers in the traps placed in a row or hedge of shrubs, than when encircled.

Allowing for the usual day-to-day variations in catch influenced by climatic and other environmental factors, mosquito reactions to these experiments were consistent enough to allow three arbitrary groups to be defined according to their responses to physical objects and levels of illumination. When taken in conjunction with information about their daytime resting places — information provided by other sampling techniques — this enabled a clearer picture to emerge about different patterns of flight path taken by the different groups of mosquito.

As an extension of these visual barrier experiments, the tub-grown shrubs of the previous series were replaced by light-weight artificial barriers, 12 ft long and 6 ft high, so constructed as to fit around the suction trap roof. The frames of the barrier were covered by nylon fishnet. Dark green cloth webbing was woven in parallel strips through each row of mesh so that two-thirds of each square was obstructed, but leaving the remaining openings sufficiently wide to permit the passage of mosquitoes. When three of these net frames were stood on their long axes, with their bases one foot apart, they gave the appearance of being almost solid, but in fact permitted free movement of mosquitoes through the mesh barrier. Three net frames were placed in line on opposite sides of the trap inlet, so that the trap fitted the gap between the ends of the six frames (Figure 2.3a). These barrier traps were set up in the same range of sites in and near the wooded swamp used in previous trials.

Mosquito response to an area of lowered light intensity was also investigated by using two large camouflaged nets constructed so as to produce shade by means of green cloth webbing strands, while at the same time allowing sufficient openings for the free passage of mosquitoes (Figure 2.3b). Results showed that in the field sites nearly all species were taken in larger numbers when the traps were furnished with net frames. In the wooded swamp, in contrast, the presence or absence of net frames had little effect on the mean number collected. When the nets were overhead, the field species were again most markedly affected, with numbers sharply reduced.

(b) Suction traps as visual objects at night

The questions raised by this first series of field experiments are of considerable importance in the context of insect response to suction traps. The main implication is that traps which were long considered to be non-attractive in theory, especially to mosquitoes and allied night-flying insects, might in fact exert some visual effect, possibly attractive in some cases and repellent in others. This important question was the object of further investigations in Florida on mosquito flight behaviour in the proximity of visually conspicuous objects (Bidlingmayer and Hem, 1979, 1981)

In the first of two experiments, two unpainted plywood suction traps that had weathered to a grey colour were used to test mosquito responses to immobile objects. The horizontal trap opening, 1.5 m above ground level, measured 79 × 79 cm, and each trap was furnished with a fan which moved 102 m^3 of air per minute. In order to determine the effect of black traps which would provide a greater contrast with the background, one

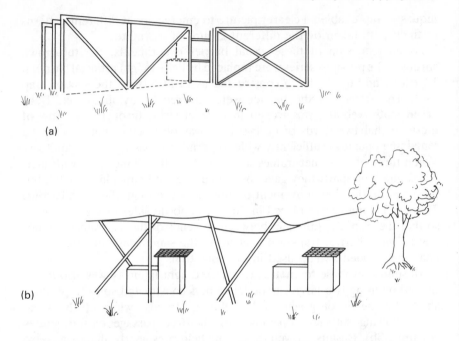

(a)

(b)

Figure 2.3 *Illustrating experiments on effects of artificial barriers on suction trap catches of mosquitoes. (a) Vertical; (b) overhead (Bidlingmayer, 1975).*

trap was covered with matt black plywood panels, except for roof, air intake and discharge. Traps were operated in two different environments, one in a salt-marsh area, and the other about 10 km inland. In addition to these two variations of the trap, a third trap with the same dimensions was designed with the object of determining the effect on catches of reduced visibility of the trap. This was done by constructing the trap of transparent acrylic plastic with the opaque material restricted to the fan, fan mounting and aluminium screen used to concentrate the insects.

In the second experiment, two suction traps without roofs were buried 40 m apart in a dyke crossing an open salt-marsh (Figure 2.4). Air was taken in at ground level and discharged at the side. Four different modifications were then tested. In the first (Figure 2.4a) the trap was provided with a rigid transparent plastic riser, 1.2 m high, with one end fitted into the trap opening. Trap collections were made at an elevation of 1.2 m with the upper end of the riser serving as an air intake. The second trap differed from the first in that the four outside surfaces of the transparent plastic riser were entirely wrapped in black cloth in order to make it clearly visible (Figure 2.4b). In the third variation only the lower 0.9 m of

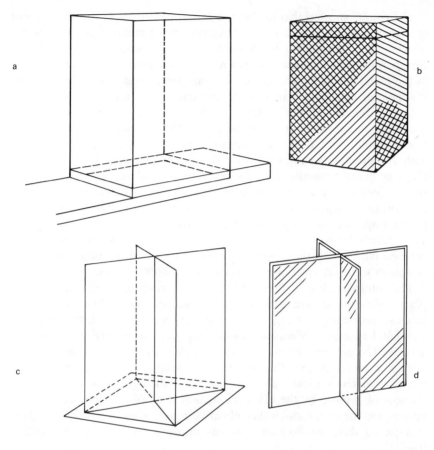

Figure 2.4 *Illustrating experiments on visual effects of suction traps. Arrangement of structures to test effects of reduced visibility (Bidlingmayer and Hem, 1979).*

the riser was wrapped in black cloth. In the fourth variation, the riser was removed and the trap was furnished with a transparent baffle formed by joining two 1.2 m high sheets of acrylic plastic at right angles (Figure 2.4c), midway along the vertical surface, the air intake now being at ground level. Within this design traps were exposed either with transparent baffles, with one baffle completely covered with black cloth (Figure 2.4d) or with no baffle at all. In order to avoid any position bias in these comparisons of visible and 'invisible' traps, trap positions were changed each day, traps being operated from sunset to sunrise.

The results showed that in the first experiment most of the 15 different species of mosquito captured were taken in larger numbers in a weathered plywood trap covered in black cloth than in an uncovered trap. When transparent traps were used, larger numbers of mosquitoes were captured in the panel-covered trap than in the transparent one. From these results it appeared that black panels were attracting more mosquitoes.

In the case of the buried traps, results showed that responses varied according to species, but for each species the reactions were consistent. For example, with *Culex nigripalpus*, a typical woodland species, traps with the completely wrapped riser and with partly wrapped riser took 2−3 times as many mosquitoes as the transparent alternative. Similarly, this species preferred black baffles to either transparent baffles or no baffles. In contrast, *Anopheles atropos* showed a consistent preference for transparent traps over those with completely covered or partly covered risers. Transparent traps were also preferred by this species over those with covered baffles and those with no baffles. Reactions of other species of mosquito were intermediate between these two extremes.

The interpretation of these differences in reaction is that in the first place visible objects are attractive to mosquitoes from a distance at night. In close proximity a change in flight direction occurs, with the adults of woodland species approaching visible objects more closely than do field species. The different behaviour patterns close to the trap, with regard to attraction or avoidance of different modifications, will also determine the extent to which the close flight pattern exposes the mosquitoes to increasing chances of capture by the effective suction zone of the fans. In order to understand the operation of this 'effective zone', more precise information is required about air flow and velocity on the approaches to the suction trap inlet.

Measurements made in still air of the air inflow 5 cm above, and horizontal to the lip of the suction trap showed velocities of 347, 238, 145, 60 and 25 cm sec^{-1} at distances of 0, 10, 20, 30 and 40 cm respectively. Laboratory experiments using flight chambers have provided data from several different sources, and with different species of *Aedes* and *Culex*, indicating that the maximum flight speed ranges roughly from 150 to 250 cm sec^{-1}. When those figures are compared with trap-intake air velocities, it appears that capture could only be certain within approximately 10 cm of the trap intake, and would decrease in effectiveness up to 20−30 cm and beyond. Thus, if certain species attracted to a visible suction trap usually pass within 30 cm of the intake, a large proportion should be captured. If other species usually pass at a distance of 30 cm or more, fewer would be captured. Consequently, visible suction traps can appear to be non-attractive, or even repellent, for those species that seldom pass within the trap's effective radius. Whether suction trap collections of mosquitoes are increased or decreased by increasing a trap's visibility depends upon both trap design and the behaviour response of different species.

(c) Mosquito reaction to suction trap discharge

While attention so far has been mainly directed to the flow of air *into* the suction trap, this same volume of air — up to $115 \, m^3 \, min^{-1}$ for each trap — is also being discharged, producing a pattern of local air disturbance. The effect of this was examined critically in a series of field experiments using standard unpainted suction traps arranged in different patterns. (Bidlingmayer and Hem, 1980). Preliminary tests in the still air of an enclosed space (an aeroplane hangar) showed that the torque of the rotating fan produced a spreading cone of discharge air which was not symmetrical but skew. At a distance of 10 m from the trap, a moving air mass was produced with a bias 4 m to one side and 2 m to the other side of the trap axis, the highest air velocity being recorded at the wider side. At 15 m the moving air mass still had a width of about 9 m, and the velocity ($0.6 \, m \, sec^{-1}$) could just be felt.

Although it is desirable in such field experiments to have comparison traps located near each other in order to ensure that trap-site factors are minimized, and that traps are exposed to equal populations of insects in flight, a compromise of 15 m apart had to be made in this series in order to minimize the air movement factor created by trap discharge. Four such traps were set up in this way in a line on an open field. The traps were mounted on a pivot, allowing them to be orientated in such a way that their discharge was perpendicular to the line or column, thus not affecting other traps (Figure 2.5a) Alternatively, they could be turned in the same direction as the line so that the two inner traps, and one of the end traps, became exposed to air streams from adjacent traps (Figure 2.5b). In each experiment therefore, depending on the orientation, one of the end traps would remain unexposed to any trap discharge.

The results showed that when the air discharge was perpendicular to the line of column, catches of females of all 12 species of mosquito involved in this test were higher in the end traps than in the centre traps. The ratio of two end traps to two centre traps was particularly high (2.6) in the case of the abundant species *Psorophora columbiae*, and in *Culiseta melanura* (1.8) and ranged between 1.5 and 1.1 for the other species.

When the traps were arranged with their air discharge in line with the column, the general picture of high catches in the two end traps versus

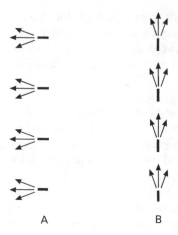

Figure 2.5 *Arrangement of suction traps to show effect of air discharge (Bidlingmayer and Hem, 1980).*

the two middle traps was repeated — with one exception. In this case the upstream end trap, i.e. the one unaffected by any other trap, recorded consistently higher captures than the downstream end trap.

In the next series of experiments a 4 × 4 grid of 16 traps, 15 m apart, was set up in such a way that trap discharge was either north or south — the results being the same in each case — in such a way as to allow trap captures to be analysed in four different ways (Figure 2.6A). This could be done by rows — traps 1–4, 5–8, etc.; columns, e.g. 1, 5, 9, 13; corner positions 1, 4, 13, 16; edge positions, traps 2, 3, 5, 8, 9, 12, 14 and 15, or inside positions, 6, 7, 10 and 11. All columns of traps had similar air flow patterns, one trap at the end of each column being unexposed to discharge, while the other three traps were exposed. One row of traps was unexposed to discharge, while the other three rows downstream were all exposed.

The arrangement also provided variation in the number of competing traps within 15 m, i.e. inner traps with four, edge traps with three and corner traps with two. The provisions of traps with only one competing adjacent trap within 15 m was made by adding four extra traps outside the grid, as in Figure 2.6B.

The experiment was therefore designed to analyse two different factors simultaneously, the effect of air stream within a group of traps, and the effect of competing visual attraction in the form of adjacent suction traps. Further modifications of this basic grid pattern were also carried out

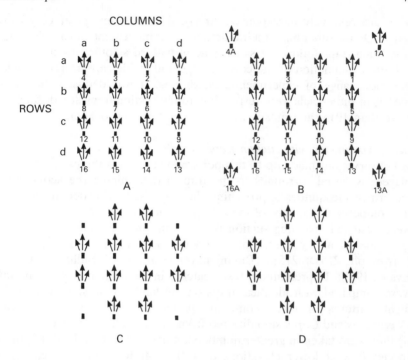

Figure 2.6 *As in Figure 2.5, Further arrangement.*

in experiments in which the four corner traps were made inoperative (Figure 2.6C) or were removed (Figure 2.6D).

The results showed first of all that while each column was exposed to a similar airflow pattern — as the traps were in line with the column — nevertheless, the exterior columns, a and d, almost invariably took larger catches, up to 50% more than the interior columns b and c. Similarly, traps in the exterior rows, a to d, captured more than the interior rows. The highest catches of most mosquitoes were recorded in the corner traps, followed by traps along the edge of the grid; the smallest numbers were taken in traps inside the grid.

For most species of mosquito in these trials, the number showed a declining trend with the increase in the number of adjacent traps, but two culicine species were apparently unaffected by the number of adjoining competing traps. The same two species differed from most others in that captures in the corner position were not noticeably greater than those recorded at the edge and inside the grid. From this and other evidence therefore it was concluded that these two species, unlike all others in the trial, are relatively unresponsive to visual targets, except at very close range. On the other hand their reactions to air movement and air discharge from such traps follow the same general pattern as other species.

From examination of the corner, edge and inside capture ratios under the wide range of experimental conditions, it was deduced that most species in these trials were responding to the suction traps at a distance of more than 10 m, and mostly in the range of 15.5 m to 19 m.

(d)   Experiments with suction traps on rafts

All observations made up to this point had confirmed the prominent part played by visual surrounds to the trap in determining the actual flight pattern of mosquitoes approaching the trap itself. In order to minimize the competitive influence of these trap surrounds, a series of experiments were designed involving suction traps set up on a floating raft. The raft was moored in the centre of an open space of water, or water-filled borrow pit, 215 × 215 m (Bidlingmayer *et al.*, 1985; Bidlingmayer and Evans, 1987). In preliminary tests, catches in the suction trap on the raft were compared with identical traps on the land. These showed that the flight patterns of three common species of *Culex* (*C. nigripalpus*, *C. erraticus* and *C. pilosus*) differed from those of all other species in that (i) they were taken in greater numbers on the raft than on land; (ii) these species flew at lower elevations over the raft than other species, and (iii) the pattern of catch indicated that culicine mosquitoes, after leaving the wind shelter in the lee of the trap, made repeated short flights into the wind, only to fall back again, each sortie increasing the risk of capture by the suction trap.

In the analysis of these flight patterns (Bidlingmayer and Evans, 1987) the plywood suction trap (6.1 × 1.2 × 1.2 m), painted flat black, was placed on the raft perpendicular to the raft's axis (Figure 2.7). The octagonal raft itself, 6.7 m in diameter, was moored by the 'bow' so that it could swing freely about its mooring, with the bow always upwind. The figure shows the location of 34 inlets in the trap, each provided with a collecting net, permitting air entry into the interior of the trap. Within the trap were two 61 cm exhaust fans, each discharging 102 m³ of air per minute. Air was drawn successively through the inlets, the collecting nets, the fan, and then discharged upwards by a curved baffle.

These experiments provided a mass of data about the mean number of mosquitoes taken per inlet per night, at different wind velocities, on the downwind side of the trap as well as on the upwind side. The inlets on the downwind side took larger catches than those on the upwind side; and inlets near the top edge of the trap captured more than at the bottom. The proportion of *Culex nigripalpus* and *C. erraticus* taken in the downwind

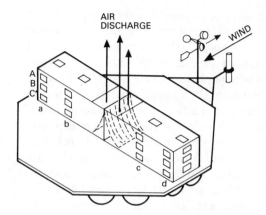

Figure 2.7 *Arrangement of suction trap on raft showing wind instruments and arrangement of air inlets into the trap (Bidlingmayer et al., 1985).*

inlets of row A were much higher than those for *C. pilosus* and the non-culicine species, indicating that the two former species were more concentrated just below the top edge of the trap than other species.

These studies showed that because of different flight behaviour among mosquito species, suction trap catches can vary greatly over short distances, even if these distances are much less than 1 m. The experiments showed wide differences in the distribution of mosquitoes captured in air inlets spaced only 0.4 to 1.2 m apart, and further highlighted the extreme importance of trap position in the accurate determination of flight paths and behaviour patterns.

2.4.2 Mosquito studies in the Gambia

The mosquito research project in the Gambia, West Africa, like the one in Florida, was equally concerned with the critical analysis of different sampling methods, particularly those suitable for detecting flight patterns of mosquitoes in the course of their regular movements between the breeding grounds in the swamps, and the villages which provide the sources of the essential blood meals. Initially, experiments were mainly concerned with ramp or interceptor traps (Chapter 6), but later suction traps, based on Vent-Axia fan units, were included in the investigation (Snow, 1975, 1976, 1979). These traps were run concurrently with the interceptor traps at farm sites for the purposes of initial comparisons. But in a later series, the traps were built out from the side of a scaffolding tower, and operated at four levels up to 9.15 m.

(a) Vertical pattern of mosquito flight

Analysis of this suction trap data enabled the mosquitoes sampled to be placed in three categories. (i) Species most common in the lowest traps, with catch declining progressively with height. This category includes several species of Anopheles, *Culex* and *Mansonia*. (ii) Species frequent at all levels, shown only by *Culex thalassius*. (iii) Species most common at the highest levels, in particular *Culex neavei* and *C. weschei*, which followed a pattern already observed in the interceptor trap experiments.

The results of these trials made it possible for the first time to define vertical flight patterns of African mosquitoes in this particular environment, namely open savanna and farmland country as distinct from the forest environment on which so much work had been done in the past. But possibly even more significant from the point of view of interpreting mosquito capture data in terms of normal behaviour patterns were the results obtained by comparing two capture techniques operating overnight under similar conditions. In the case of *Anopheles melas* for example, the brackish-water breeder and main vector of malaria in this area, both capture techniques revealed the same vertical flight pattern, with largest catches at the lowest levels and smallest at the highest level. However, the total number taken in the flight or interceptor traps were trivial compared with the large catches taken in the suction traps, providing strong evidence that this species was avoiding the flight traps.

In the case of *Culex thalassius*, more mosquitoes were taken in suction traps than in flight traps in the two years when these two techniques operated concurrently. This further illustrates the important conclusion that there are major differences in the efficiency of these two trapping methods for different species of mosquito. Further proof of this was provided when the results of comparison tests were considered separately according to whether the moon was present or absent. These figures showed that the general pattern of vertical distribution of most species was for the most part uninfluenced by the presence or absence of moonlight, but the size of the catch was strongly affected. For example, in the case of *Culex thalassius* and *C. decens*, suction trap catches appeared to be uninfluenced by the presence or absence of moonlight, but the flight trap catches on the other hand showed much higher numbers recorded under moonlit conditions, indicating that these 'non-attractant' traps are in fact visually attractive to these species in the presence of moonlight.

Of particular interest at this stage of the programme was the introduction of a new sampling device in the form of a directional suction trap (Snow, 1977). This is based on the conventional, and non-directional, Vent-Axia suction trap used in the earlier studies, but with an important modification; this took the form of a green netting funnel with a wide entrance, 1.2 ×

1.2 m, into which mosquito flight is channelled (Figure 2.8). The netting mesh was wide enough to have minimal effect on wind flow round the traps, nor did the fans themselves affect this flow. Traps were arranged in pairs, and at two or three levels, with the funnel of one pointing upwind and one pointing downwind. The captures obtained confirmed the general pattern of upwind flight as revealed by the directional flight traps noted above, except insofar as *Anopheles melas* was concerned. With that species, the directional suction traps indicated predominantly upwind flight at all levels. This appears likely to be a more accurate picture than that suggested by captures with the interceptor, ramp-type trap, to which this species is visually responsive.

In the first series of experiments on study of vertical distribution of mosquitoes by means of suction traps attached at various levels to a scaffolding tower, it emerged that with some mosquitoes the maximum density occurred below the lowest level sampled, i.e. 0.68 m. Accordingly, a second series of trials was carried out in open savanna near tidal brackish swamps which were the main breeding grounds of the two dominant mosquito species, *Anopheles melas* and *Culex thalassius* (Snow, 1982). In addition to the previous levels sampled, 1.0, 2.1, 3.9 and 7.9 m, four

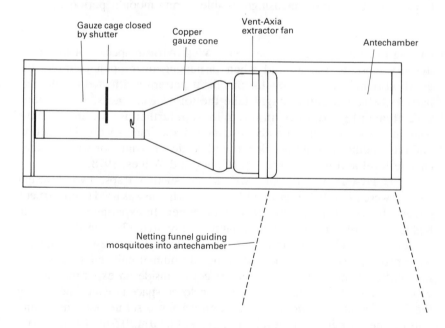

Figure 2.8 *Plan of directional suction trap used in the Gambia (Snow, 1977).*

lower levels were sampled with suction traps fitted with 22.8 cm Vent-Axia extraction fans. These were sited over pits excavated at the corners of the tower base at levels of 1.0, 0.5, 0.25 and 0.1 m. Observations were also made on wind speed, time of capture, and presence or absence of moonlight.

The data on vertical distribution now enabled four categories to be defined. (i) Vertical distribution and density of mosquitoes declining progressively with height. This category included all unfed *Anopheles*, as well as gravid *An. melas*, of which over 60% were taken at or below 0.25 m. (ii) Densities increase to around 0.5 to 1.0 m, and then decrease steadily with height. This included all catches of unfed *Culex thalassius*. (iii) Frequent at all heights; a few species only. (iv) Most captures at higher traps; *Culex poicilipes* and *C. weischei*.

The continued overnight sampling showed that in both the dominant species, *Anopheles melas* and *Culex thalassius*, the flight altitude declined progressively during the night. The presence or absence of moonlight had no demonstrable effect on the flight altitude of either of these species, but with *A. melas* there was a reduction in numbers captured during moonlit periods, which was more marked at high illumination levels. This was taken as evidence of an avoiding reaction on the part of this species to the traps as they became increasingly visible during moonlit periods.

(b)   Effect of physical barriers
Having established the fact that many West African species of mosquito fly very close to the ground when dispersing over open country, the question remained as to how far this flight pattern is affected by physical barriers, which in nature might take the form of tall grass, regenerating vegetation or belts of trees. In this follow-up, further use was made of the column of suction traps already described. Experiments were designed to find out whether mosquitoes flew round vertical barriers, or whether they change level and fly over the top (Gillies and Wilkes, 1978)

In the lay-out of artificial barriers and suction traps, two different designs were used. In Experiment 1 black polythene was used to construct a circular fence 2.9 m high and 18 m in diameter. In experiment 2, a much higher fence of 6 m surrounded a large grassy area 130 m in diameter.

Independent checks on mosquito density throughout these experiments were provided by catches of mosquitoes on human bait, either sitting in the centre of the smaller enclosed space, or inside an experimental hut constructed in the centre of the larger enclosed space in experiment 2. In the first experiment, a column of suction traps was set up both inside and outside the barrier, at six levels, i.e. ground or 0 m, 0.5 m, 1.0 m, 1.5 m, 2.0 m and 3 m high. Collections of *Anopheles* were much the same inside and outside the barrier, but in the case of the dominant culicine, *Mansonia*, densities were significantly lower in the centre of the enclosed area, compared with the open ground outside. The suction trap catches disclosed that with *Mansonia* catches at ground level, i.e. zero height, were three times greater outside the fence than at the same level inside, but from the level of 0.5 m upwards, differences were not significant. These results indicated that the immediate effect of introducing an obstruction into the flight path of *Mansonia* mosquitoes was to produce a different pattern of vertical stratification inside the fenced area.

In the second experiment, three columns of suction traps were set up just outside the fence at distances of 0.25, 1.0 and 13 m. At each site the suction traps operated at three levels, 0.5, 1.0 and 5 m above ground level. The object of this arrangement was to study the mosquito flight pattern in the critical approaches to the high fence, and compare it with the normal pattern observed in the absence of a fence. The results obtained by the human bait collection in the experimental hut, supplemented by light trap collections made automatically over a calf placed in a mosquito-proof stall inside the hut, showed that the presence of the high fence had no effect on the relative numbers of *Mansonia* and *Anopheles* in the hut. However, the catches in the array of suction traps just outside the fence disclosed that the two groups of mosquito reacted very differently to the barrier obstacle. In the low-level suction traps close to the fence there was a significant increase in the density of low-flying *Mansonia*, as in Experiment 1, compared with the normal pattern, but in traps 13 m from the fence there was a significant reduction in density of *Mansonia* in the presence of the fence, indicating that mosquitoes were accumulating at low levels close to the fence; this caused a relative fall in density, both in the higher traps and in the traps more distant from the fence.

In view of the fact that mosquito densities inside this large enclosure — as judged by captures in the experimental hut — were not reduced by the presence of the fence, mosquitoes must therefore have entered the area over the top of the fence. In the case of *Mansonia*, the effect of the fence on its normal low flight path was to lead to an accumulation of mosquitoes at low levels close to the fence. Eventually these mosquitoes scale the fence and enter the enclosed area without any fall in density. Within the larger fenced area with the diameter of 130 m there was evidently sufficient space for the normal low flight of *Mansonia* to be quickly resumed, as judged by the sustained levels of catches in the experimental hut. Within the smaller fenced area (diameter 36 m) it appears however that there is not sufficient space for the normal low flight pattern to be resumed, an effect which is indicated not only by a reduction in numbers captured on the host at ground level, but also by the alteration in vertical stratification revealed by the suction traps inside the fenced area, which recorded catches at ground level as being much smaller inside the fence than outside.

2.4.3 Suction traps in woodland mosquito studies

The suction trap technique has been found particularly useful in two quite different situations, both involving woodland mosquitoes of temperate regions, one in the USA and one in England.

(a) Studies in England
The woodland mosquito species of England, of little economic or public health interest, attracted very little scientific attention until the advent of myxomatosis in the rabbit populations in the mid 1950s, when *Aedes cantans* and *A. annulipes* were shown to be efficient vectors of the virus, in the laboratory at least. Apart from their blood-feeding attacks on humans entering these woodlands, which provided the ideal breeding pools, practically nothing was known about their population structure, feeding habits and normal diurnal movements.

Intensive studies involving suction traps initiated in the late 1960s, changed the situation completely with regard to both the anopheline and culicine species concerned (Service, 1971). Studies on flight periodicity and vertical distribution were carried out mainly by means of this technique. Aerial populations were sampled initially by Johnson−Taylor suction traps, and latterly by Vent-Axia traps, sunk into the ground so that the fan openings were at a height of 30 cm (Figure 2.9). These operated over 24 hours and automatically segregated the catches into hourly collections. These ground level catches were supplemented by suction traps operated initially at heights of 23, 73, 123, 173 and 223 cm, and latterly up to 550 cm. The traps were arranged so that expelled air from the fan was delivered to the side of the trap, and not directly downwards where it might have disturbed the air intake of the trap below. The smooth intake and operation of fans was checked by smoke tests.

In all species, including *Aedes cantans*, unfed females constituted more than 96% of the total catch, with males normally 1% or less. Blood fed and gravid females made up less than 2% of the catch. These suction trap catches were supplemented by 10 human bait collections from 16.00 to 22.00 hours. These independent sets of data enabled flight periodicity to be compared with the biting cycle, and showed that with *Aedes cantans* the time of arrival of hungry unfed females at bait was similar to the flight times of unfed females as shown by the suction traps, with a peak from

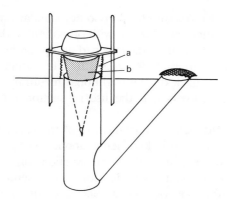

Figure 2.9 *Suction trap sunk into the ground for sampling woodland mosquitoes. a, Flexible tubing; b, fine wire mesh supporting the cone (Service, 1971).*

19.00 to 20.00 hours. The suction trap experiment showed that females of *Aedes cantans* — as well as several other species investigated — which feed mainly on mammals, fly very close to the ground at 30 cm or less, with less than 3% recorded at heights in excess of 100 cm.

(b) US studies on virus vectors
In the USA, in contrast to the UK, there is a long history of critical studies on the ecology of indigenous mosquitoes, mainly due to the fact that so many of them are of economic or public health importance. But the woodland mosquitoes themselves attracted comparatively little attention until recent years when the potential of two of the common species as vectors of virus was established, namely *Aedes sierrensis* in the pacific north west and California as a potential vector of Western Equine Encephalitis (WEE), and *Aedes triseriatus*, incriminated as a major vector to man of Californian encephalitis virus in mid-western USA, as well as being a likely vector of La Crosse virus in Wisconsin.

Intensive investigations by one team from Wisconsin from 1967 onwards, supplemented by a further team, from Indiana, from 1974 onwards, has produced a wealth of information on the ecology of these mosquitoes. However, the main interest in the context of this chapter is the comparatively recent introduction of the suction trap technique into this research programme, with particular reference to *Aedes triseriatus* which for long had presented difficulties in the way of sampling adult populations. *Aedes triseriatus* is mainly a day-time biter, and is not normally taken in light traps. It also proved refractory to many other standard mosquito capture methods such as animal-baited traps, with or without added dry ice (CO_2) (Sinsko and Craig, 1979). Investigators had no option but to make use of the fact that in study areas *Aedes triseriatus* bites humans at a relatively high rate, and to fall back on the traditional recording of landing/biting rates on human volunteers and the use of suction tubes or aspirators (Beier *et al.*, 1982).

As in England, the belated introduction of the suction trap technique resulted in a rapid increase in knowledge about flight patterns and population densities (Novak *et al.*, 1981). The suction trap used by the team from Notre Dame, Indiana, was developed independently of the UK design (Figure 2.10). The trap was an enclosed cone type made up of a 25 cm diameter sheet metal cylinder, with a sheet metal elbow flared up to a 58 cm diameter opening. The flared opening was painted black with contrasting white stripes to provide a discrete visual attraction. Below the elbow was a screen concentrating zone, with a collecting unit mounted at its base. The 28 cm diameter suction fan was mounted below the collecting

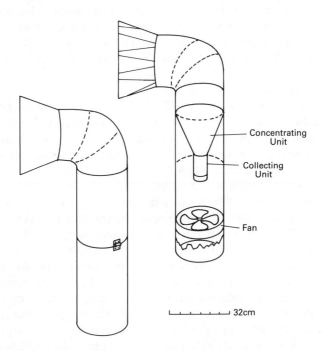

Figure 2.10 *Enclosed cone type of suction trap used for woodland mosquitoes in the USA (Novak* et al., *1981).*

unit, thus ensuring that insects did not have to pass through the fan blades where they might have been damaged.

Suction traps were set up at different levels from ground (1.8 to 2 m) to canopy (27–31 m). In addition to target species, a total of 20 species of mosquito were captured. Both males and females of *Aedes triseriatus* were taken in great numbers in the canopy. One species captured abundantly, particularly in the canopy, *Anopheles barberi*, had only rarely been collected by any of the other capture techniques used prior to this study, i.e. by human bait, light trap or by CO_2 fitted traps. The investigations also revealed an additional complicating factor in interpreting sampling data, namely that *Aedes triseriatus* exhibits diurnal biting activity at ground level, and crepuscular activity at 21 m, a point of considerable significance in transmission of virus from wild hosts to man.

2.5 Suction trap experiments with blackflies (*Simulium*)

The design and application of suction traps for insects of medical and veterinary importance such as mosquitoes and biting midges has been strongly influenced by the pioneer work established at Rothamsted on agricultural pests. Nevertheless, studies on these two disciplines have rarely overlapped or integrated, with one notable exception, namely the possible use of this technique for sampling the small biting blackflies (*Simulium*) which form the target of the World Health Organization (WHO) onchocerciasis control project in West Africa, initiated in 1976 and scheduled for 20 years.

Savanna members of the *Simulium damnosum* complex — the vectors of human onchocerciasis or 'river blindness' — now appear to be migratory species, capable of long-range movements, and responsible for the 're-invasion' of hitherto controlled areas. Initial disappointments with the suction trap technique for *Simulium* relegated it to limbo for many years, but new demands for sampling techniques to supplement the standard catch of female flies attracted to human bait, and expressed as number of flies per boy h^{-1}, led to its revival and re-examination. But what provided the unusual stimulus was that one of the main participants, and consultant, in this research team was one of the pioneer workers responsible for the early development of suction traps and trap catch evaluation at Rothamsted Experimental Station in England. The combined expertise of this team, representing specialists from medical and from agricultural fields of entomology, provided an unusually good opportunity for a critical examination of this technique (Johnson *et al.*, 1982).

The siting of traps in the West African savanna test ground had to be made according to the availability of mains electric supply, as well as to adequate populations of blackflies. Ten traps mounted on wooden stands so that their inlets were 3 m above ground were operated continuously day and night. Two kinds of trap were used, a 45 cm (18 in) propellor and a 30 cm (12 in) aerofoil fan, both of the enclosed cone type (Johnson and Taylor, 1955). Each propeller trap sampled approximately $73.6\,m^3\,min^{-1}$ $(2600\,ft^3\,min^{-1})$ and each aerofoil trap about $36.8\,m^3\,min^{-1}$ $(1300\,ft^3\,min^{-1})$. At one location a pair of propeller traps were used, and at another four aerofoil traps. At each location the group of traps sampled $212\,000\,m^3$ in 24 hours.

The suction traps were very successful in one direction in that over 9000 female *Simulium* were captured, including rare species not previously recorded. But in another direction, the work failed in its main objective in that only a single specimen of the target species group, *Simulium damnosum*, was recovered. Another common species, the non-target *Simulium hargreavesi*, was also completely absent from these traps even though it was breeding abundantly in local streams.

In all these West African catches there was also an almost complete absence of any but unfed females; the absence of engorged females removed any hope that this sampling technique would provide ample material for the identification of blood meals, and a much-needed clue to the natural feeding habits of these blackflies on hosts other than man.

This inexplicable difference in reaction of different species of *Simulium* to suction traps, as well as to light traps, had been established in earlier trapping tests in Scotland (Davies and Williams, 1962; Williams, 1962, 1965). Light trap catches recorded an abundance of two species, *Simulium tuberosum* and *S. 'latipes'* which bred abundantly in local streams, but the catch in a suction trap operated nightly nearby at the same time was found to be composed almost entirely of the latter species alone.

Similar differences in reaction to such traps among different species of *Simulium* have also been reported more recently from Canada (Shipp, 1985) in a comparison between several different trapping methods, silhouette, sticky and suction, used both with and without the added attractant of dry ice. In this case the suction trap was simply a New Jersey light trap (I, 2.2) without the bulb. Among the four most abundant species tested, the reactions of *Simulium aureum* — a mainly ornithophilic species — differed sharply from those of the other three in that the suction trap caught more than any other method, 51% of the total compared with 20% *S. arcticum* and only 9% *S. decorum*. The addition of the CO_2 bait to the suction trap increased the catch of *S. arcticum* by seven times and of *S. aureum* by four times.

2.6 Remote sensing devices to check trap efficiency

2.6.1 Suction traps

Comparative tests described earlier in this chapter defined the 'efficiency' of suction traps by comparing them with other trapping devices such as aerofoil traps, rotary traps and 'tow' nets. From this and related work it emerged that four primary factors or parameters enter into this concept of efficiency. These are the average wind speed in $m\,sec^{-1}$, insect size in mm, the volume of suction rate in $m^3\,sec^{-1}$, and suction trap inlet diameter. One of the clearest relationships among these factors is the demonstration that trap efficiency decreases with increasing wind speed, and also with increasing insect size.

This decreasing effect in turn could be due to other factors involved, either acting singly or in combination. For example it might be caused by insect trajectories which instead of entering the stream 'tubes' converging on the trap inlet, cross these suction lines and thus miss the trap inlet. This in turn might be influenced by the insects' flying speed on approaching the trap. Or again, there might be a rapid change in insect flight behaviour upon sensing the acceleration of air stream near the air tubes.

These are features which cannot be clarified by the indirect methods used so far. What is needed is more precise information based on the detection of insects actually approaching the trap inlet, and the measurement of the fraction captured in the trap inlet air stream. For this purpose the possibilities of two remote sensing devices have been explored. One of these is by means of infra-red detection using a system called IRADIT, i.e. infra-red active detection and automatic determination of insect trajectories (Schaefer and Bent, 1984). The other possibility which has been explored even more thoroughly is the use of Radar (Schaefer *et al.*, 1985), with particular reference to the sampling of aphid populations by the Rothamsted type trap.

Radar has proved to be of value for larger insects ranging in size from the desert locust (5 cm) — to the spruce budworm moth (1 cm). As insect size decreases, the echo intensity falls, and is marginal for insects of aphid size, i.e. 0.2 cm, but can be used when there is a large outbreak of mainly one species. By the use of Radar it could be shown that during each suction trap trapping period of 30 minutes, the trap sucked in $1440\,m^3$ of air, but the effectiveness was only from 0.3 to 0.6, say 0.4. This means that an aerial density of one insect in $600\,m^3$ would produce on average one trapped insect per half hour sampling period. Even at this low density the Radar pulse volume of $1800\,m^3$ would contain about three similar insects continuously.

The Radar pulse samples air passing through a vertical cross-section of, say, 11×15 m, and would create a sample rate near $500 \, \text{m}^3 \text{sec}^{-1}$, which is some 1200 times the volume sampling rate of a standard Rothamsted trap. A 2 minute Radar record represents many more aphids than would be caught in a 30 minute suction trapping.

These techniques confirm the decreasing trap effectiveness with increasing wind speed, and this may mean that over a period about 50% of the insects may be lost from the sample. This could be important early in the immigration season when only one specimen is captured every few days. In the Rothamsted insect survey it is the usual practice to sum the catches over 7 day periods for use in aphid forecasts, but during that period there are considerable fluctuations in wind speed, and hence daily trap effectiveness will vary considerably. Wind speed often has a diurnal pattern with maximum speeds in the early afternoon.

2.6.2 Light traps

Radar has also proved to be a valuable tool for the study of moth movements in relation to light traps. It is a natural extension of direct viewing of moth flight, providing a remote sensing technique capable of detecting flight paths when these are no longer visible to the human eye, either because of distance or of failing light.

A combination of light traps, direct viewing of moth flight from a high tower and radar played an important role in determining the dispersal pattern of egg-carrying females of the spruce budworm, *Choristoneura fumiferana* — often over considerable distances — which leads to periodic outbreaks and heavy defoliating infestation (Greenbank *et al.*, 1980). The same combination has been used to good effects in Australian studies on the flight of *Heliothis* moth species which are major pests of cotton. Visual observations on moth flight above crops were made with binoculars and assisted by a spotlight. In this way insect take-off could be observed visually at a height of a few metres, but with radar-assisted detection this range could be extended to 2 km (Biggin *et al.*, 1986).

The radar was able to detect the existence of a plume of moths downwind of the trap. This plume, which extended for a distance of about 200 m, and which was sometimes as broad as it was long, was observed dissipating and reforming when the light was switched off and on. The flight of these moths took them repeatedly towards and then away from the light. Most of the moths turned away from the light at distances of a few metres, and the number eventually trapped in the light trap was relatively small.

Other radar observations showed that, with moths flying over light traps, flight is disrupted up to 400 m from the trap, and that, in fact, very little attraction to the light occurs (Farrow and Daly, 1987). However, it has not yet been possible to determine the extent to which moths ascending or descending through the boundary layer respond to light traps.

Chapter 3

Pheromone-Based and Sex Lure Traps

3.1 Introduction

The powerful attraction exerted by female moths over males, often operating over long distances, has long intrigued entomologists, particularly those concerned with the monitoring and control of insect pests. However, it is only within the last twenty years that this subject has undergone considerable development, to the extent that it has dominated, and continues to dominate, the whole field of insect response to trapping. Attempts to use sex attraction as a practical means of trapping moths, particularly in the USA date back to well before that period. In the case of the gypsy moth, *Lymantria dispar*, traps baited with female genitalia were used as far back as 1932 (Steck and Bailey, 1978). Traps using virgin females to attract males were also used in the early 1960s for trapping codling moth, *Cydia pomonella*. Traps baited with virgin females of the spruce budworm, *Choristoneura fumiferana*, were used in New Brunswick, Canada, from 1960−1967(Miller, 1971)

The period of intensive research and development from the late 1960s was triggered firstly by the identification of specific sex-attractant chemicals or pheromones in several moth species, and, secondly, by the synthesis of an increasing number of these pheromones in the laboratory (Weatherston *et al.*, 1971). Following that work, sex attractants have largely replaced many traditional monitoring methods for most pest species of moth.

The great potential of this new technique for monitoring, or even controlling, populations of moths of agricultural importance led to a great proliferation in the development and field-testing of pheromone based traps. The increased practical possibilities also resulted in a range of commercially produced traps, all of which had to be tested against the many different pest species concerned. The variety and intensity of trap testing in the 1970s has provided material of the greatest interest, both in the experimental procedures used in field testing and the use of parallel laboratory studies in order to extend or clarify various aspects of insect response to pheromone based traps.

One important factor in all this experimental work, and in the great advance in knowledge, is that it is only comparatively rarely that the specific sex pheromone has turned out to be a single chemical compound. One of these exceptions is the gypsy moth, *Porthetria dispar*, in which there is a single chemical compound, reproducible in the form of the synthetic pheromone Disparlure But in the majority of species, the sex pheromone is composed of more than one identifiable compound, often as many as three or four, and up to six. Each of these components may play a different role at the different stages between the initial attraction of the male, and those leading to the final mating (Cardé and Charlton, 1984). As a consequence of this, the whole question of insect response to pheromones has become extremely complex, and this complexity naturally affects the evaluation of capture data from pheromone based traps.

The development and application of pheromone-based traps over the last twenty years falls into two phases. The first of these is the intensive development, design and field testing of various types of pheromone trap against a wide range of economically important moth species. This was the phase dictated by immediate practical considerations, in the confident belief that the development of synthetic pheromones and their incorporation in suitable traps would provide an answer to many problems in the monitoring and control of these pests. The second phase, and one which logically might have preceded the first, rather than follow it, is marked by even more critical studies on all aspects of insect response to pheromones. This has been increasingly investigated as a problem of pure science, requiring the specialized experience of insect physiologists familiar with the design of appropriate laboratory equipment for analysing insect responses. While work on the first phase, evaluation of pheromone traps in the field, is now easing off gradually, the second phase continues to gather momentum, with hundreds of new reports published each year. This book will concentrate on a selection of moth species which have received particular attention over the years, and which best illustrate progress in the design, use and evaluation of pheromone based traps.

3.2 Development of pheromone-based traps for the spruce budworm *Choristoneura fumiferana*

The spruce budworm is a widespread forest pest in many parts of eastern Canada and north-eastern USA. Due to wide population fluctuations, periodic upsurges lead to severe outbreaks causing extensive defoliation. Methods for monitoring adult population changes of this moth, sensitive enough to forecast potential outbreaks, have long been unsatisfactory. Lack of standard methods has made it difficult to compare results from different areas. Prior to the advent of pheromone, conventional methods for assessing abundance of adult moths, such as light traps and mark-release techniques, proved to be of limited value. Emphasis was on the sampling of immature stages, namely larvae, pupae and egg masses (Morris, 1955). This proved to be a very exacting process in which aluminium pole pruners, extension ladders, tree trestles and platforms were all necessary to obtain samples of foliage from tall trees.

3.2.1 Earlier experience with sex traps

Prior to the identification and synthesis of spruce budworm pheromone in 1971 (Weatherston *et al.*, 1971, Sanders and Weatherston, 1976), live virgin females themselves formed the attractant in sex-based traps. These were used in south-west New Brunswick, Canada, from 1960–1967 (Miller, 1971; Miller and McDougall, 1973).

The sex trap was a single board 10 × 24 × 0.5 in, with a hole drilled about 10 in from one edge in which was inserted a plastic tube (1.5 × 3 in long) which formed the cage for an unmated female, both ends of the tube being covered with mesh. The board was smeared with tanglefoot on both sides. Three trap configurations were used: (i) 25 traps 20 m apart, with each trap fastened to a tree trunk at eye level; (ii) ten traps 20 m apart along a trail at eye level, and (iii) larger traps, 10 × 48 in, with two caged females, fastened to dominant trees at the mean height of the codominant stratum. Five of these 'crown level' traps were placed 100 m apart. Traps were operated for about 20 days in any one season, beginning with the onset of male emergence. Ground level traps were checked daily, male moths counted and dead females removed; crown level traps were checked weekly.

Catches were found to be very variable, often at the level of a single moth per trap. With the lowest mean seasonal count per trap recorded at 2.0, the indications were that the 25 traps used for 20 nights were inadequate, and that 80–100 traps would have been required to provide figures of comparable value. Trap efficiency was also found to be affected by the nature of the adhesive material, and moths did not always stick at first contact. By means of traps set at six different levels, from 0.5 to 45 feet on a 45 ft tower in a dense stand of balsam fir, it was found that six times as many moths were taken within the crown stratum as at ground level.

Observations were also made on the age of the females used in these experiments, and the results indicated that they were most attractive on the fourth night. Females of the spruce budworm moth start 'calling', i.e. releasing pheromone, a few hours before sunset, and it is at this time that the first males are trapped with activity reaching a peak about two hours after sunset.

One method of testing 'trap efficiency' was by comparing moth counts with larval counts within the same forest plot. Changes in the ratio between these two measurements clearly showed that the trapping effectiveness of a virgin female is affected by population density. For example, at a resident density of 70 males per acre, the female attraction was ten times higher than when the male density was 4000 per acre.

3.2.2   Identification and development of sex pheromones and trap design

With the advent of synthetic sex attractants or pheromones, there was a great intensification of work on trap design and trap performance from the early 1970s onwards. A dominant role was played in these trials by the new commerciably available traps whose performance had to be compared for a variety of species and under a wide range of conditions. Such brand names as Pherocon and Sectar appear with increasing frequency in these reports. In one of the first series of trials against spruce budworm, only four of these commercial traps were available, and these were subjected to critical evaluation in the field (Sanders, 1978, 1988) (Figure 3.1). Pherocon IC is simply a cardboard trap covered inside with adhesive, and a cardboard cover leaving an opening 2 cm wide all round. Pherocon ICP is the same except that the bottom trap fits tightly under the cover, except for two entrances at the opposite sides. Sectar I trap has adhesive on all inside surfaces, while XC-26 is a larger version of Sectar I, requiring a wooden cross member to prevent it collapsing. In all traps the bait was PVC formulation of synthetic sex attractant, each trap being baited with a pellet 4 mm × 10 mm long.

Trap performance was evaluated in spruce plantations in areas of high defoliation, but some trapping was also carried out in areas where there had been no outbreaks for 20 years. Although it was known that the highest catches of spruce budworm moth occurred in the upper canopy of

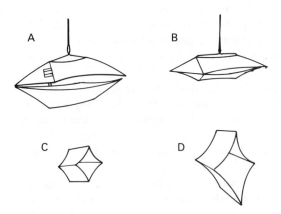

Figure 3.1 *Commercial pheromone traps used in field trials on spruce budworm. A. Pherocon IC; B. Pherocon ICP; C. Sectar I; D. XC 26. (Sanders, 1978, 1988).*

host trees, for practical purposes traps were all set at a height of 2 m. In evaluating these trap catches in relation to total moth population, the long-established larval count methods were to prove very useful. For example, in 1975 it was found that densities of 4th and 5th instar larvae were 0.4 per 45 cm of branch tip, which corresponds to a few thousand adults per hectare, a density at which traps could prove most useful.

While the value of these new sex attractant traps for monitoring moth densities at low population levels soon became well established, a complication was found to affect estimates at high population densities. At those levels the ability of traps to capture insects declined as the catch increased due to (a) coating of the trapping surface with debris, twigs, pine needles, etc.; (b) reduction in the effective area of trapping surface as more males were captured; and (c) repellent effect of insects already captured. The decline in effectiveness in traps exposed for two weeks was compared with the catch in fresh traps (Table 3.1). Using a Sectar I trap it was found that after 14 days the average total catch of males in the unchanged trap was 49.3 compared with a cumulative catch of 225.7 in the fresh trap. In the XC-26 trap, the corresponding figures were 87.8 and 353.8. In a further experiment it was found that traps with the remains of trapped males were less attractive due to olfactory repellence.

Using a grid layout for the comparison of the different trap designs, Pherocon IC proved to have the highest capacity, followed by ICP and XC-26, with Sectar I as a close fourth. These differences were found to apply at both high and low moth population densities. In this series the

Table 3.1 Comparison of catches of male *Choristoneura fumiferana* in two types of trap, Sectar I and XC−26, left unchanged for a 14-day period(A), and freshly baited(B) traps, changed every 2 days.

	Ratio A/B × 100	
Date	Sectar I	XC-26
July 6	104	116
July 8	79	78
July 10	43	77
July 12	32	13
July 14	14	25
July 16	11	9
July 18	8	14

(After Sanders, 1978)

superiority of the Pherocon IC trap was also evident in its weathering ability over a 12 month period, which is evidently determined by the protection afforded to trapping surfaces. This has also been found with other species of moth, such as the fruit tree leafroller, the grapeberry moth and the tuberworm moth.

3.2.3   Trap saturation, trap density and trap age

The problem of variables associated with pheromone trapping for monitoring spruce budworm moth populations was further studied in an area of high density in Maine, USA, and low population densities in Ontario, Canada (Houseweart *et al.*, 1981). The experimental area in Maine had been sprayed with the insecticide Sevin in 1976, but not in 1977 and 1978 when the tests were carried out at periods spanning peak flight activity of moths in July. The emphasis was on those variables which affect trap efficiency, namely, 'trap saturation', trap density or spacing and trap age. Pherocon traps hung at 2 m and baited with synthetic attractant, Fulure, were tested under a variety of conditions: (i) baited trap with unchanged bottom; (ii) baited trap with fresh bottom every 2 days; (iii) baited trap with an aged bottom; and (iv) unbaited with an unchanged bottom as control.

The results showed that the number of moths caught per trap was approaching saturation when it reached the 50–60 level. Traps do not cease to operate at that level, but the rate of capture substantially declines. In outbreak centres in Maine it was found that Pherocon traps could functionally saturate overnight, and that at high population levels this saturation point was sometimes attained within an hour. The highest moth catches were in traps where the saturation factor was reduced by replacing the bottom every day. As the bottoms of traps became older, fewer males were caught, presumably because the sticky surface became less sticky. As trapping efficiency diminishes with age it follows that traps deployed several days before moth emergence may be less effective than traps dispersed during the flight period.

It has become clear from this and other investigations (McLeod and Starratt, 1978; Starratt and Mcleod, 1976) that the problems of trap saturation and trap age pose serious obstacles in trying to establish any reliable correlation between catches and population density. In order to counter this effect, the potency of the pheromone lure could be reduced so that not more than 50 males are attracted between collections. Alternatively, the bottoms could be changed frequently before the critical 50 moth level is reached, or a non-saturating, high-capacity, no exit trap could be designed.

This last approach was adopted by designing a trap which dispensed with the sticky trapping surface, replacing it with a liquid catching surface (Kendall *et al.*, 1982). This 'Kendall' trap (Figure 3.2) has a capacity of approximately half a litre and is provided with four entry holes 3 cm diameter. A 3:1 mixture of 70% ethanol and ethylene glycol is added to each collection jar as a killer-preservative fluid. With this design the maximum catch for a single trap/night was 1197 males. The trap was also found to record much higher catches than the original bucket trap (Moser and Brown, 1978) on which it was based.

3.2.4 Trap efficiency

The problem of trap saturation was encountered with many other moth species in trials elsewhere. In the case of the spruce budworm, the question was re-examined at a time when the value of sex pheromone traps for monitoring long-term population changes had become firmly

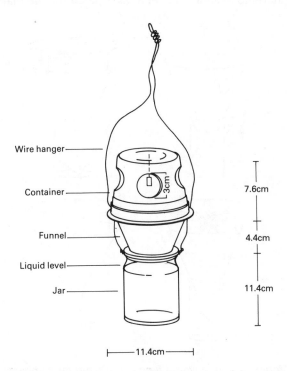

Figure 3.2 *Kendall pheromone trap designed to replace sticky trap surface with liquid catching surface (Kendall* et al., *1982).*

established, in a programme utilizing about 10 000 traps yearly in the adjoining regions of Ontario, Canada and Michigan, USA (Ramaswamy and Cardé, 1982). These trials were concerned with a comparison of conventional sticky traps and various 'non-saturating' traps, all baited with formulated sex lure. Traps were hung 10 m apart at a height of 2 m, and the trials involved various modifications of the Pherocon IC traps with the bottoms replaced every day. The trap designs were mainly covered funnel traps, or CFTs, having openings of different sizes and configurations (Figure 3.3). Dichlorfos was used as the killing agent. CFTs which utilized entrance ports, irrespective of size, number or location (Figure 3.3a) were not found to be as efficient as those which employed an opening all round, between the top plate and the funnel rim (Figure 3.3, b–d).

Some features which were found to contribute to the success of the best of the traps tested were: (i) an opening between the top cover and the rim of the funnel to ensure that the trap is omnidirectional; (ii) a screen cage around the pheromone source to serve as a perch for the males; and (iii)

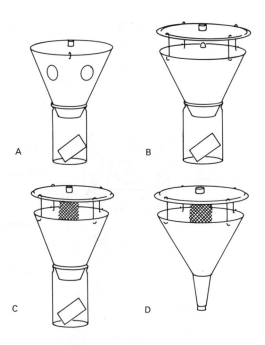

Figure 3.3 *Modification of covered funnel trap used for comparison of various 'sticky' traps and 'non-saturating' traps baited with sex lure (Ramaswamy and Cardé, 1982).*

dichlorvos close to the pheromone source in order to achieve quick knockdown. A comparison was then made between the best of the CFTs and the standard commercial Pherocon IC. One set of each was left intact for 24 hours, and in the other set the Pherocon traps had their bottoms replaced after ten moths had been caught, while the CFTs were cleared of trapped males every hour. The results showed that Pherocon traps left untouched for 24 hours were only 57% as effective as those which had the bottom replaced as soon as ten moths had been captured, the daily means being 89 and 156 moths respectively. In contrast, the CFTs left untouched for 24 hours captured a daily mean of 64, nearly twice the number caught (33) in traps checked every hour. This apparent anomaly was attributed to the fact that in the untouched CFTs the insecticide had more time to act on trapped moths, while in the traps checked hourly, some moths were disturbed and escaped before the insecticide had time to act. It was sometimes observed that as many moths flew out of the trap as were killed in it, and it was concluded that dichlorvos was not the ideal killing agent.

While the sticky surfaces of the standard Pherocon IC are likely to lead to trap saturation in untouched traps, overall the performance of the CFTs was only about 60% that of Pherocon. Nevertheless, these trials showed the potential of non-saturating CFTs in providing an alternative to sticky traps for monitoring spruce budworm moth populations.

3.2.5 Assessment of pheromone trapping for spruce budworm

Information on spruce budworm control was provided by a three-year (1981–1983) field assessment of the role of pheromone-baited traps, at sparse to medium budworm population densities, based on a collaborative effort between four Canadian provinces and four north-eastern states of the USA (Allen *et al.*, 1986). Four trap models were used (Figure 3.4): (i) A CFT (covered-funnel) trap with dichlorvos (see page 102) with slight modification to abate water collection; (ii) Pherocon ICP; (iii) and (iv) two commercially available gypsy moth traps. The lure in these traps took the form of a PVC pellet containing 0.03% concentration of pheromone, this concentration approximating to that of a 'calling' female spruce budworm moth. Baited traps, hung at 2 m, were placed in the field four days before the initial flight period and retrieved seven weeks later.

The results showed that in all plots the Pherocon ICP quickly became saturated. Among other findings emerging from these trials was the fact that 'lure inconsistency' was considered a major factor in determining variation in trap catch. The role of 'edge effect' in affecting trap performance within spruce stands, as compared with results obtained in traps

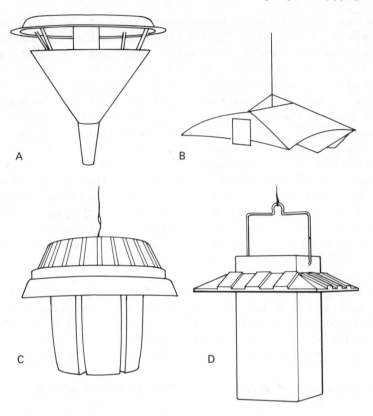

Figure 3.4 *Pheromone-baited traps used in collaborative Canadian/US trials*
A. CFT design; B. Pherocon ICP; C. GMPC (Gypsy moth trap); D. GMMC
(Gypsy moth trap) (Allen et al., 1986).

sited along the edge of stands, was also demonstrated. The full value of
trap monitoring carried out at 2 m above ground level is restricted by the
fact that budworm egg masses, and consequently major adult activity, is
concentrated at the upper crown level where routine setting of traps
would be impracticable.

Progress in the development of four pheromone traps over a period of
21 years is well illustrated by the sequence of events in one location in
north-western Ontario (Sanders, 1988).

1966–71   Period before identification of sex pheromones. Traps based
          on virgin females and sticky boards.

1972–76 Application of sex pheromone traps. Sectar I traps used baited with polyethylene caps containing pheromone lure.

1977–86 Polyvinyl chloride lures containing pheromone were used.

1977–80 Pherocon IC traps used.

1981–86 High-capacity non-saturating traps used. Double funnel traps in the first half of the period, and International Pheromone System 'Uni-traps' in 1984–86.

The monitoring of population trends of the spruce budworm over that period was based on both pheromone trap records and on densities of late instar larvae. Comparison of the two methods (Figure 3.5) shows that a significant increase in larval density in the late 1960s was mirrored by the trap catch. Moth populations reached outbreak status in 1983, with extensive defoliation. Moth movement may occur over many tens of kilometres, and can thus give trap catch figures higher than the local population levels. As methods for estimating egg and larval densities are very inaccurate, particularly at low population densities, a coordinated programme has been established in eastern North America for the annual deployment of sex pheromone traps to monitor spruce budworm populations. By 1986, these included 550 localities spaced over seven Canadian provinces and six American states.

The second aspect of the work on the spruce budworm concerns the experimental approach to trap response and moth behaviour. While field

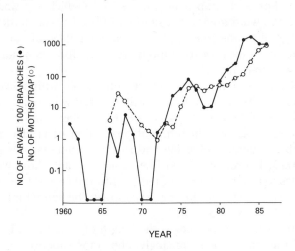

Figure 3.5 *Comparison of population trends of spruce budworm moths in Ontario over 21 years, by pheromone trap records (– – – – –) and by densities of late instar larvae (———) (Sanders, 1988).*

observations involving a comparison of different types of trap under similar conditions can provide the first valuable guide to general levels of trap efficiency, it soon became necessary to examine the term 'efficiency' more critically. The term is usually interpreted as 'the number of insects approaching a trap that enter and are retained by the trap'. Information on this point can only be obtained by experimental methods, either by release of known numbers of moths in the vicinity of the trap, or by wind tunnels, or by both.

The value of the wind tunnel technique had already been demonstrated in the case of the western spruce budworm, *Choristoneura occidentalis* (Angerilli and McLean, 1984), before it was applied to the true or eastern spruce budworm, *Choristoneura fumiferana* (Sanders, 1986). The wind tunnel used in the latter was 90 × 90 cm in cross-section, and 190 cm long, in which the wind speed was adjusted to approximately 25 cm sec^{-1}. Male spruce budworm moths were released individually about 1 m downwind of the trap, and the times at which moths entered, left and re-entered the trap were recorded. Observations were made over a period of five minutes from release.

Among many factors affecting trap efficiency is the structure of the pheromone plume in the trap vicinity (Lewis and Macauley, 1976; see page 130). This was measured by dropping titanium chloride onto pheromone lure and photographing the cloud (Sanders, 1986). Seven types of non-saturating traps plus two sticky traps were used in this experiment (Figure 3.6). For each trap a measure of trap efficiency was made by recording, in the wind tunnel, the number of moths (standardized at 50), the percentage of moths entering traps, the percentage of entering moths retained, the percentage of total moths retained for 5 min, and the time of first entry (Table 3.2).

The results showed that the highest entry rates within five minutes, ranging from 86–90%, all occurred in those traps which had unrestricted openings with omnidirectional access, i.e. Multi-pher, Uni-trap, CFT and Pherocon ICP. Efficiency, in terms of proportion of moths captured in relation both to numbers entering and to total numbers, was greatest in the sticky traps, Pherocon IC and Pherocon ICP, and those with funnel-shaped baffles to prevent escape of moths, Multipher and Uni-trap, which retained 91–93% of moths entering. In contrast, the figure for the Covered Funnel Trap (CFT) was only 51%.

Another measure of efficiency is the length of time taken to locate the entrance of the trap. Not surprisingly, the traps superior in this aspect were those with the widest entrances — CFT and Pherocon IC — with times of first entry 0.13 and 0.11 minutes respectively. Ease of access can also be offset by ease of escape, and this was evident in the CFT where

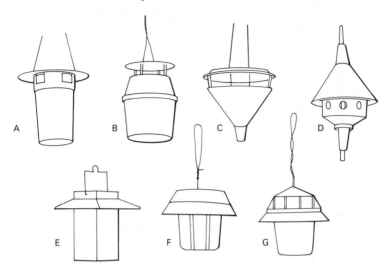

Figure 3.6 *Pheromone traps used to determine structure of pheromone plume as a measure of trap efficiency. A, Multi-pher; B, Uni-trap; C, Covered funnel trap (CFT); D, Double funnel trap; E, Milk carton trap; F, Gypsy moth trap; G, Bag-a-bug canister (Sanders, 1986).*

half the moths entering the trap escaped before being overcome by the insecticide.

3.2.6 Pheromone blends, lure potency and male response

The question of lure potency is also an increasingly important factor in trap evaluation. In the case of the spruce budworm, the main component of the sex pheromone is (E:Z)-11-tetradecenal of which a 95:5 blend provided adequate catch of male moths, even though additional minor components are also present in natural pheromone (Sanders and Weatherston, 1976). The question of lure ageing was tested by removing lures from storage each week, beginning six weeks before the anticipated start of male flight, and pinning these in Pherocon ICP traps left out for one week, or for less that two days in high density areas, and checking the traps daily (Sanders and Meighen, 1987). The trial also included a comparison of five formulations of lure.

The results showed the importance of maintaining lure concentrations as close as possible to that naturally released by virgin females. Concentrations in excess of this were liable to cause abnormal male behaviour,

Table 3.2 Efficiency of sex pheromone traps in a wind tunnel as indicated by the percentage of male spruce budworm moths that are attracted to the traps, enter them, and are retained for 5 minutes, and the average time taken for the moths to enter the traps.

Trap design.	Number of moths.	% moths entering trap	% of moths entering retained	% of total moths retained.	Time of first entry. (Min)
Multi-Pher	50	90	91	82	0.21
Uni-trap	50	88	93	82	0.31
Covered funnel trap	50	86	51	44	0.13
Double funnel trap	50	50	76	38	0.33
Gypsy moth 'milk-carton'	50	68	29	20	0.30
Gypsy moth canister	50	44	23	10	0.43
Bag-a-Bug	50	16	62	11	0.23
Pherocon IC	30	87	100	87	0.11
Pherocon ICP	30	67	90	60	0.25
Pherocon ICP (oriented crosswind)	30	73	86	63	0.32

(Sanders, 1986).

such as arrestment of upwind flight, leading to reduction in catch. Some lures produced catches which were too low and could fail to detect low populations, while others proved to be too attractive, making counting of captured moths difficult at high densities. The vital importance of combining all these variable factors to produce maximum trap efficiency can be better appreciated when it is borne in mind that pheromone trapping is only effective for a very short period (three weeks) of the year in which actual flight of spruce budworm moths takes place, any variations in which are allowed for by having an actual trapping period of six weeks.

It is also clear from these and other observations that response of male spruce budworm moths to calling females is still significantly faster, and can be maintained longer, than response to synthetic blends of pheromone, indicating that the synthetic blend is incomplete (Sanders, 1984; Silk and Kuenen; Silk *et al.*, 1988, 1989). Another factor emerging is the mating status of captured males. Both mated and virgin male *Choristoneura* are recorded at pheromone-baited traps (Bergh *et al.*, 1986), and a complete evaluation of trap response must take these two categories into consideration.

The mating status of male moths can be determined in many species by the nature and presence or absence of fluid transferred from male to female during copulation. If this is pigmented, its absence from the male indicates that it has mated. Attempts were made to determine whether differences exist between the response of mated and unmated males to pheromone traps, or if they had different flight paths which would affect their capture. This was done by setting a pheromone-baited trap on a scaffold tower at three different elevations, from 2.5 m to 11 m, corresponding to bottom, middle and top of the forest canopy (Bergh *et al.*, 1988). The number of males increased from bottom to top of the canopy, but the proportion of mated males was independent of elevation. However, it was suggested that future critical evaluation of trap response by spruce budworm moths should divide the male moth catch into three categories: (i) males devoid of coloured secretion, i.e. those which had mated within the previous 24 hours; (ii) males with a pale-coloured secretion, indicating that mating had occurred 24−48 hours previously; and (iii) those with a dark secretion, indicating virginity.

3.3 Pheromone research: Select examples of trapping experience

By the early 1980s the continued intense interest in pheromones had resulted in the chemical identification of sex pheromones of more than 100 species of moth, totalling more than 200 compounds (Bjostad and

Roelofs, 1983). In most of these cases identification of sex pheromones was followed by a series of experiments on trap design and trap response, comparing sticky commercial models of trap with non-sticky, non-saturating models under a range of conditions. A few select examples will be considered here.

3.3.1 Spruce budmoth: overlap of problems with spruce budworm

Spruce budmoth larvae, *Zeiraphera canadensis*, feed primarily on shoots of white spruce. In 1980 this species created a problem in New Brunswick, Canada, when it became extremely abundant. This led to an intensification of research into the role of pheromones and pheromone-baited traps. These outbreaks occurred in the same region in which spruce budworm, *Choristoneura fumiferana*, was a long-established forestry pest. The presence of both species in the same area produced unusual complications in evaluating pheromone trapping.

Sampling methods for spruce budmoth, like those for spruce budworm, originally depended on time-consuming procedures involving the separation of eggs and larvae from the foliage, and counting them. The development of a synthetic pheromone for budmoth led to extensive field trials with the various compounds concerned (Z and E isomers of dodecenyl, tetradecenyl and hexadecenyl acetates and alcohols) (Turgeon and Grant, 1988). Tree heights in the oldest white spruce plantations were 2.5 m, allowing pheromone traps to be hung in the upper crown of selected trees in a way which was not feasible with the much taller trees infested with spruce budworm. The Pherocon ICP traps which were used were arranged in a grid pattern 25 m apart, and had their two openings reduced in size from 5.5×4 cm to 1×1 cm with a fibreglass screen in order to lessen the number of spruce budworm moths which entered these traps, presumably by accident. Similar reductions in trap funnel orifice from the normal 3 cm to 1 cm were made in the tests involving Uni-traps.

The pheromone was species-specific to budmoth, and the fact that invasion of these traps by budworm was accidental was borne out by the finding that budworm invaded pheromone-baited and unbaited traps in equal numbers when its density was highest. At high densities there was also a tendency for budmoth itself to blunder into unbaited controls, up to 84 males being recorded in one trap.

Trials were also carried out to compare trap performance at three levels, middle crown 0.75 to 1.0 m, upper third of crown 1.5 to 2.0 m, and a third level 0.5 m above the apical bud of the leader, sampled by means of an extended pole. The upper crown traps were found to catch more than those either above or below. The attraction of the crown level was further emphasized by recording high catches of moths in unbaited traps in the canopy. It was suggested that the visual stimuli provided by tree shape or silhouette apparently helps males to orientate in search of females.

3.3.2 Experiments with the gypsy moth, *Lymantria (Porthetria) dispar*

(a) Trials on trapping
Two unusual features of gypsy moth biology make it ideally suited for experimentation on response to pheromone-baited traps. First, the natural pheromone consists of a single chemical compound, which was been synthesized to produce a commercial lure, Disparlure. The second feature is that this woodland and orchard species is active by day, thus allowing its trap responses to be observed in the field as well as in the laboratory (Cardé and Charlton, 1984). This species also has a long history of trapping by means of sex lures, as traps baited with female genitalia were used as far back as 1932 (Steck and Bailey, 1978; Collins and Potts, 1932)

In epidemic outbreaks of gypsy moth, densities may exceed 10000 hectare^{-1}. The females emerge from the pupae from mid-morning to late afternoon, and initiate 'calling' from sites near ground level to the top of the tree canopy within an hour of eclosion. Males also emerge throughout the same daily period, and undertake vertical flight-and-search for the females.

Early in the pheromone trap studies, six kinds of cone trap were compared with two commercial pheromone sticky traps, with open access (Figure 3.7) (Steck and Bailey, 1978). Models 1−4 had two identical cones per trap; models 5−6 had a single cone and were wind directional on a swivel spike; models 7−8 were Pherocon coated cardboard designs. In experiments involving four different moth species, three of them noctuids, it was found that the commercial sticky traps, mainly designed for orchard use, were of limited value in the Great Plains environment in which these particular tests were carried out. These traps were found to be incapable of retaining more than 50−60 noctuid moths of average size, and were liable to the saturation effect experienced with the spruce budworm experiments. In contrast the cone model traps could retain 200 moths in the case of models 1 and 2, while 3, 4 and 5 had a capacity of 7500 moths. However, the open access sticky traps did have a practical role to play when moth populations were low, or when low potency lures were used; under those conditions they proved to be more efficient than cone traps. The large wind-directional model 5 was found to be consistently good, but there was no clear 'best' design among 1−5.

These tests also examined the effect of trap orifice, which should ideally allow maximum ingress with minimal egress. It was found that the best effect was achieved in traps with an orifice diameter about double

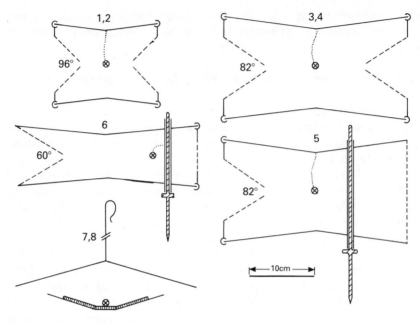

Cross-sectional plans of trap designs. Lure dispensers
indicated by ⊗, copper attachment wires by dotted lines.

Figure 3.7 *Range of pheromone traps used in field trials on gypsy moth. See text for guide (Steck and Bailey, 1978).*

the thorax width of the target species (Figure 3.8). With cone holes having a diameter 2.5 to 3.5 times greater than the thorax width, there was no retention of males.

In the earlier phases of pheromone studies it was believed that the pheromone of each moth species was a single chemical compound which stimulated successive steps in the sequence of moth behaviour with increasing concentration. However, it is now known that the majority of moth pheromones are multicomponent, except the gypsy moth which conforms to the original simplistic concept. The fact that this pheromone is a single component which can be synthetized makes this species an ideal subject for analysis of response, and this question has been closely examined in laboratory studies (Cardé, 1979)

(b)   Behavioural response in wind tunnels
The whole natural process of pheromone release by the female, known as 'calling', may be continuous in some species, while in others it may take

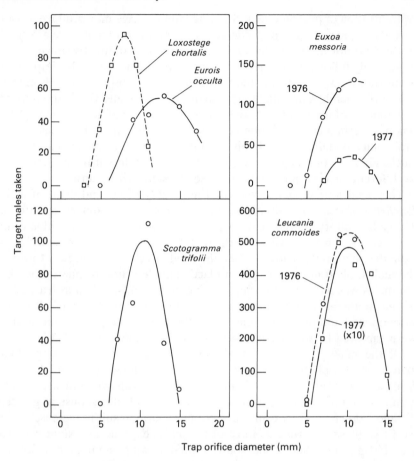

Figure 3.8 *Relation between pheromone trap orifice and ingress/egress for five moth species (Steck and Bailey, 1978).*

place in periodic bouts or pulses. In this connection the continuous emission of synthetic pheromone from trap dispensers may not relate directly to the natural pulsed emission. Wind tunnel experiments have an important part to play in defining male behaviour responses in the early sequence of events, although they have limitations in analysing subsequent flight reaction in the field.

In the laboratory the sequence of behaviour in the wind tunnel is as follows: (i) wing fanning; (ii) pre-flight walking; (iii) up-wind flight; (iv) landing or flight arrestment; and (v) courtship behaviour. Decreasing concentration of pheromone at the wing fanning stage may produce longer periods before response, and a decreasing proportion of males are affected. In the field upwind flight does not normally proceed directly upwind, but more likely follows a zig-zag course about the plume axis. Also important is the duration of upwind flight, which can be observed in laboratory tests. As to landing and flight arrestment, in most males this is the normal culmination of upwind flight. In some species with multicomponent pheromones, certain chemical members may have the effect of increasing the probability of landing, and this usually leads in turn to increased trap catch. Finally, courtship behaviour, takes many different forms according to different species, and there is no set or standard pattern.

This behaviour sequence was used in a study of upwind flight of male *Porthetria dispar* towards a pheromone source (Elkinton *et al.*, 1987). Thirty laboratory reared male gypsy moths were placed individually in mesh cages, 1.6 m above ground, at distances of 20, 40, 80 and 120 m from the pheromone source. When the wind direction was directly towards the males, pheromone was put in place. The detection of pheromone by the males was indicated by wing fanning movements. When at least 30% of the moths at all four distances had initiated wing fanning, all cages were opened and males approaching within 2 m of the pheromone source were captured by insect net and their time of capture recorded.

Of the males that flew out of the mesh cages, the percentage that arrived at the pheromone source within the following 40 min declined from 45% at 20 m to 8% at 120 m. In field tests with synthetic attractant, high release rates or more attractive blends may elicit response at greater distances down wind, but this does not necessarily result in increased trap catch. In these experiments most of the males exposed at distances of 80 and 120 m from pheromone source initiated wing fanning, but few of these flew to the site of pheromone release over the 40 min interval.

(c) Pattern of pheromone plume
In order to define the pattern of the pheromone plume from the point of release, puffs of mineral and smoke from a generator were tracked by an observer, as were 7 cm diameter neutrally buoyant, helium-filled balloons underneath a forest canopy which is the natural environment of the gypsy moth. At 5 s intervals, the position of the balloons or the smoke puff was marked by a stake in the ground. These observations showed that the trajectories of these two indicators were often highly non-linear, even over distances less than 10 m (Figure 3.9). This non-linear trajectory of pheromone puffs in the forest experimental area suggests that moths flying upwind in response to pheromone will in many cases not head directly towards the source. The most important effect of increased wind

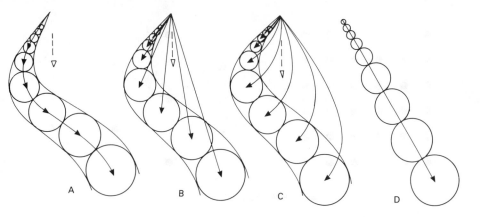

Figure 3.9 *Patterns of pheromone plume from point of release as indicated by puffs of mineral and smoke (Elkinton et al., 1987).*

speed on pheromone communication is to decrease the meander in wind direction, resulting in longer tracks of successive puffs travelling in the same direction. Thus for gypsy moth, the effective distance of communication, and the number attracted to the source, are likely to increase with wind speed, at the lower end of the range of wind speeds at which upwind flight occurs.

The observation that gypsy moth males can detect Disparlure from commonly used sex-attractant traps at a distance of at least 80 m led to a further investigation on the effect of inter-trap distance and wind direction (Elkinton and Cardé, 1988). This study was concerned with the possibility that the number of males captured at one pheromone trap might be influenced by the presence of other traps less than 80 m distant. In one experiment, traps were deployed in a 6 × 6 grid system, with an inter-trap distance of 80 m, while in another, groups of seven traps were deployed in a hexagonal array with inter-trap distances of 2.5, 5, 10, 20, 40 and 80 m (Figure 3.10). The interpretation of the results of both experiments was affected by the finding that the number of males captured per trap was higher round the perimeter of both grid and hexagon than in traps in the centre. The hexagon trap showed that this perimeter effect occurred at all inter-trap distances, except at 80 m, the central traps catching significantly fewer males than the mean number in the six perimeter traps. The obvious explanation of this perimeter effect is that those traps were the first encountered by moths approaching from outside the array.

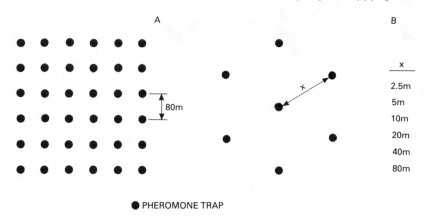

Figure 3.10 *Arrangement of pheromone traps in experiment on effect of inter-trap distance and wind direction (Elkinton and Cardé, 1988).*

(d)   Studies in the UK on the flight paths to pheromone source
The unusually suitable features of *Lymantria* as an experimental subject
for basic research on pheromones has been further exploited by research
workers in England, a country remote from the natural zoogeographical
region of this species. This was achieved by importing into UK under
licence gypsy moth pupae supplied by the US Department of Agriculture.
This research was aimed at studying the track of the pheromone plume in
the field, and simultaneous flight track of the male moth (David *et al.*,
1983). Direct observations on flight were only made possible by the day-
time habits of this species; in order to make these more easily visible, the
wings of male moths were dusted with pink fluorescent powder while they
were temporarily immobilized.

An area of $400\,m^2$ in the field was marked in a grid every $2\,m$ with
numbered poles. The attractant, Disparlure, evaporated from $30\,cm^2$ filter
paper was placed near the upwind end of the area, $30\,cm$ downwind and
$10\,cm$ above the orifice of the bubble generator. The generator produced
a stream of detergent bubbles, each about $5\,cm$ in diameter. Moths were
exposed singly in an open cage near the downwind end of the experimental
area, in or near the plume of bubbles when they took off. Each moth was
flown up to five times through the area, being recaptured by butterfly net
each time and released. The bubbles had no effect on the behaviour of
the moths.

Video records of the plume of bubbles showed that the pattern conformed
to previous conclusions with smoke plumes (David *et al.*, 1982) and
exactly mirrored the variations and fluctuations due to changes in wind

pattern. Of 31 moths that took off on release, 22 reached the odour source, doing so in 72 out of 130 flights, flying mostly at 0.5 to 1 m above the ground. When moths were observed to be flying among bubbles — which indicated the odour track — the pheromone source always lay within 15° of due upwind of the moth, indicating that the moth was actually heading towards the source with not more than 15° error. When away from the bubbles, the moths hardly ever advanced straight upwind, and in 68% of cases flew to and fro across wind on wider casting excursions. When a moth lost contact with a plume of bubbles because a change of wind direction had carried away the next portion of the plume, it stopped advancing upwind, and cast across the new wind direction. This was done in such a way that when the casting moth re-entered the plume, it usually did so at a point closer to the source than the point where the plume had been lost.

3.3.3 Codling moth (*Cydia pomonella*): effect of trap density and seasonal changes in trapping efficiency

The use of sex attractants for monitoring codling moth populations in orchards dates back to the early use of traps baited with live virgin females (Vakenti and Madsen, 1976). The codling moth pheromone was discovered in 1970 (Roelfs *et al.*, 1971), and since that time has largely replaced traditional monitoring methods. The main emphasis has been on the application of synthetic pheromones to pest management in order to provide information to growers on levels of population at which infestation might occur (Howell, 1972; Westigard and Graves, 1976). A great deal of this practical work has involved the development of non-sticky traps to counter the well established saturation element in sticky traps (Howell, 1984; Vakenti and Madsen, 1976). A new design of trap, the cup trap (Figure 3.11), using a liquid retainer in the form of polypropylene glycol was found to overcome many problems of accumulating debris inherent in pheromone traps with sticky surfaces (Howell, 1984).

Catches from these various traps all point to the low population levels which can still be significant in pest monitoring. In one investigation, the trap in question, the Batiste trap (Westigard and Graves, 1976) captured an average of 1.6 males per week, while in another, based on the commercial Pherocon trap (Vakenti and Madsen, 1976), it was concluded that if codling moth densities averaged two or more per trap per week, growers were advised to spray.

On the question of trap spacing and trap density, early experiments showed that the number of codling moths per trap increased between a density range of 1 trap/50 trees to 1 trap/1500 trees (Riedl *et al.*, 1976).

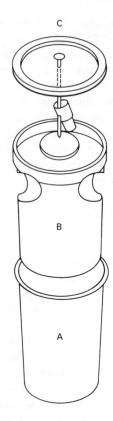

Figure 3.11 *Cup-trap design of pheromone trap used in gypsy moth studies. A, 5-oz paper cup; B, 6-oz styrofoam cup with holes cut in side; C, Plastic lid with pheromone-bearing septum suspended from it. (Howell, 1984).*

In a further investigation it was confirmed that the increase in the number of traps in relation to tree density will increase the total catch, but reduce the number caught per trap (McNally and Barnes, 1981) (Table 3.3).

An unusual situation was studied in Michigan where there are two generations of codling moth. Here, pheromone traps were found to be more effective at the spring flight, but trapping efficiency declined towards the end of the first generation, and was generally lower during summer flights (Reidl *et al.*, 1976). Implicit in the use of pheromone traps are possible competition and interaction with the female population in the area. Because of this there is a question as to whether the same interpretation can be put on trap catches throughout the season. This competition

Table 3.3 Trap catch of codling moth in relation to tree density (all trees at 2 m)

Trap/Tree density	Relative number of moths caught per trap (Relative to lowest = 1)
1 trap per 50 trees	10
1 trap per 32 trees	9.4
2 traps per 32 trees	5.5
4 traps per 32 trees	3.5
8 traps per 32 trees	2.5
16 traps per 32 trees	1.8
32 traps per 32 trees	1.0

for males between traps and feral females depends on the ratio of female density — which is variable — to trap density, on whether the sex ratio is different from unity, and on the relative attraction of the pheromone source. In this particular case, male codling moths predominate in the first spring emergence, but later in the season increasing population and domination by the females leads to reduced trap efficiency.

3.3.4 Oriental fruit moth (*Grapholitha molesta*): studies on pheromones and on pheromone trapping in Australia and the USA

The oriental fruit moth is an international pest which has long posed problems in the way of devising standard methods for monitoring populations. In areas of its distribution as widely separated as the USA and Australia, it has been the experience that light traps do not normally capture this species. The existence of pheromone of oriental fruit moth was demonstrated in 1965 and the identity of the main component established in 1969 (Roelofs *et al.*, 1969).

(a) Australian trap comparisons
With light trapping ruled out it has long been the practice to use baits such as wine, molasses and sugar solutions to trap oriental fruit moth. In Australia, bait pails have been widely used but have proved difficult to maintain due to evaporation in hot weather, and attraction of insects other than the target moth. However, this was the obvious technique with which to compare tests on pheromone-based traps (Rothschild *et al.*, 1984). It was found that in both spring and summer flights, daily collections at both kinds of trap fluctuated greatly. There was also considerable fluctuation in the sex ratio of moths captured at bait pails. In this investigation therefore, it was not possible to obtain meaningful estimates of the absolute density of adult moths in the field, against which to assess the

catches in bait pails and pheromone traps. There was also no correlation between the catches by these two methods and the results of band-trapping emerging adults. No evidence could be established for any relationship between the numbers and proportion of catch of females captured daily in bait pails, and the number of males taken in pheromone traps.

Trials with different design of commercial and other designs of trap showed that trap catches were not appreciably affected by trap shape and dimension, although in general catches were higher in models with the largest basal area and width (Rothschild and Minks, 1977). No significant difference was noted in the relative number of males landing on discs treated with pheromone, with or without decoys. There was therefore no evidence that the presence of other moths leads to the differences noted between traps cleared regularly and those serviced only at 3 week intervals. In observations on the effect of quantity of pheromone per trap, it was found that after 21 days of exposure there was no significant difference between traps baited with concentrations ranging from $10-1000\,\mu g$; but few males were taken at concentrations below $1\,\mu g$.

(b)   Direct observations on approach of males to pheromone traps in the field

As a counterbalance to these discouraging results, Australian workers have devised research methods of considerable importance to pheromone studies in general. Field tests were carried out using both caged live females as attractant, in triangular traps with the flat base coated with adhesive, and traps with pheromone released from closed polyethylene caps or from stainless steel planchettes. All traps were hung at 2 m high (Rothschild and Minks, 1974). It was noted that in all traps, only relatively few of the males approaching the traps were actually captured; accordingly the emphasis concentrated on the approach pattern of moth flight. By means of a succession of observers on successive nights detailed information was obtained on the timing of daily flight activity in relation to traps. It was observed that moths flew upwind towards the pheromone source, and made crosswind casting movements when within 1 or 2 m of the pheromone source. In flights other than the spring emergence, male activity began at from 2 h to 30 min before sunset, and reached a peak at from 1 h before to shortly after sunset (Figure 3.12). The flight times of males at traps containing live virgin females and those baited with synthetic pheromone were similar. This happens despite the fact that while the synthetic pheromone is released continuously, that released by the live female is only emitted at particular periods, as it is with other tortricid moths as well as the codling moth.

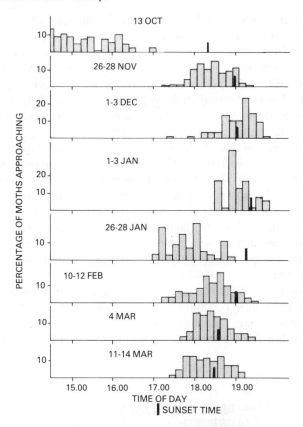

Figure 3.12 *Flight activity of male oriental fruit moth in relation to traps baited with pheromone and virgin females showing seasonal trends (Rothschild and Minks, 1974).*

(c) Observations in the USA on effectiveness or 'drawing' range of traps
Associated with the flight pattern of male moths approaching traps is the problem of the effective or 'drawing' range of traps. This is a matter of great importance in pest management; for example, if traps attract moths from outside the crop area, they could overestimate the number of females within the crop in relation to infestation danger. These aspects of trap response of insects, relevant to many general problems of evaluating pheromone trap catch, have been closely examined in the oriental fruit moth in the USA (Baker and Roelofs, 1981). In the experimental plots used, pheromone was released at the rate of 1, 10, 100 and 1000 µg from

loaded rubber septa placed on top of a 1.7 m pole. A smoke plume generator was used to indicate wind direction, and a separate marker for pheromone plume position. Using three caged males at a time, the cage was taken at walking pace from a distance of several hundred metres downwind, towards the pheromone source. The cage had an open end so that activated males, stimulated by pheromone, could fly unhampered towards the pheromone source. With the aid of fluorescent markers, behaviour patterns were monitored all the way.

The results recorded the mean maximum distances of response initiation at which males first exhibit walking, fanning while walking and upwind flight, and were found to vary directly with the dosage at source (Figure 3.13). This pattern was retained despite considerable day-to-day variations, caused mainly by the effect of temperature variations on male response. For example, no males took flight below 16°C. Also, upwind flight was totally suppressed at high wind velocities. At the highest dosage, 1000 µg, walking followed quickly by upwind flight was initiated about 80 m from pheromone source.

Observations on flight of single males towards the pheromone source enabled the termination of flight to be observed. Although the two higher release rates had longer average maximum effective distances than lower rates, the higher rates active distances did not extend all the way to the

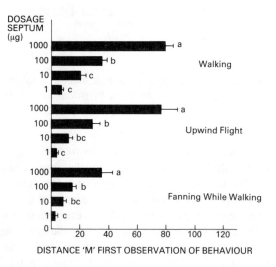

Figure 3.13 *Mean maximum active distances of walking, upwind flight and fanning exhibited by oriental fruit moth while walking to four pheromone dosages (Baker and Roelofs, 1981).*

source. A higher percentage of males terminated upwind flight, before reaching the source, to the 1000 µg and 100 µg, although just as many males initiated flight as to the 10 µg septum.

The average upwind active distances became skewed away from the source as release rates increased. It appears therefore that increasing the drawing range beyond a certain limit may be offset by a reduction in numbers of males-in-flight which actually reach the attractant source. The result of this is a decrease in trap efficiency in the males captured divided by the number of approaches. The average termination distances from the attractant source at the two concentrations of 1000 µg and 100 µg, were 155 and 20 cm respectively. These observations are supported by those of other workers who have also found that, despite the greater drawing range, traps containing 1000 µg capture fewer males than those with 200 µg (Roelofs and Cardé, 1977; Roelofs, 1978).

(d) Male response to blend/dosage combinations and effects of temperature

The effects of temperature on male response to pheromone, responsible for day-to-day variations in male flight activity, were examined in both the oriental fruit moth and the pink bollworm (Linn *et al.*, 1988a) Those experiments were not limited to the main pheromone component, but included a range of combinations with two other components. It was already known that oriental fruit moth exhibits its peak level response to a relatively narrow range of blend/dosage combinations of the chemicals. The indications from this were that there are two threshold effects on male moth flight activity: (i) inability of males, once they have taken to flight, to orientate to the odour plume and begin upwind flight; this was exhibited at lower dosages or blends; (ii) males exhibit arrested flight to the odour signal once they have initiated upwind flight; this occurs with higher blends of pheromone components than occur naturally. A slightly different response was shown by the pink bollworm moth which showed peak level of response to a wider range of treatments, with no evidence of an upper dose or blend arrestment threshold.

Experiments with wind tunnels have now shown that the specific response of these two species to pheromone can be altered dramatically by temperature. The responses of oriental fruit moth to different blends and dosages at 20° and 26°C have been charted for taking flight, upwind flight and source contact. The results show that previously established differences in male responses of these two species were due to temperature effects on male perception of pheromone, and not to inherent specific differences in male sensitivity. Male behaviour is most affected by temperature at the stage of plume orientation and initiation of upwind flight, and it appears that this temperature effect is in no way due simply to increased rate of pheromone release. These experiments show that temperature can have a dramatic effect on male perception of multi-component pheromones, but leave unresolved the extent to which this affects mating success in the field (Linn *et al.*, 1988b).

3.3.5   Pea moth (*Cydia nigricana*): experimental studies in the UK on range of attraction of traps

Many of the observations on range of attraction of pheromone traps described so far have been made in forest or orchard environments in the USA. In marked contrast to this is the environment of the pea moth which has been the subject of so much inventive experimental work in the UK (Wall and Perry, 1978, 1980, 1983, 1987; Perry and Wall, 1984). The site of this work was an open wheatfield which formed an emergence site for pea moths which had bred in the pea crop previously grown on the site. This work re-examined the problems involved in laying out field experiments with pheromone-baited traps and unbaited controls. The 'trapping zone' was considered as delimiting the volume within which individuals respond sufficiently to be trapped. It is important to establish the extent to which this zone is affected by competition or interference from other traps in the vicinity. As interaction between traps can be significant even at 100 m, this factor may well invalidate experiments involving too high a density of traps.

(a)   Comparison of single trap catch with flanked trap catch
The basis of these trials (Wall and Perry, 1978) was to compare first, the number of moths caught in solitary traps with similar traps surrounded or flanked by a circle of eight other traps, arranged at different spacings, such as at radii of 5, 15, 25 and 100 m. All traps were placed just below crop height. Secondly, the experimental layout took the form of a line of three traps in which the central trap was compared with the two outer traps, arranged either along or across wind (Figure 3.14). The results of the first trials showed that the effect of the surrounding circle of traps was to reduce the central trap catch by 88% when the radius was 5 m, and 37% when the radius was 100 m, showing that mutual interference between traps is still operative at that maximum range tested.

In the experiment involving traps set along the wind direction, the upwind trap caught more moths than the other two, at both spacings tested, the order of catch being upwind > central = downwind. Figure 3.14b shows diagramatically that a trap which is downwind, and within the trapping zone of another, can catch moths originating from the central

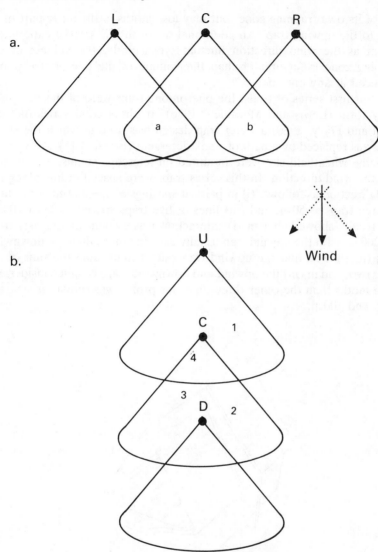

Figure 3.14 *Arrangement of pheromone baited traps across wind (a) and along wind (b) in experiments with pea moth (L, left; C, centre; R, right; U, upwind; D, downwind) (Wall and Perry, 1978).*

part of its own trapping zone, but may lose moths in the outer part of its zone to the upwind trap. An additional factor in this kind of experiment is that as the wind direction fluctuates constantly, the volume of the trapping zone is far greater than the volume of the pheromone plume produced at any one time.

In this first series of trials, the pheromone traps were of the triangular sticky form (Lewis and Macauley, 1976). In the second series of trials (Wall and Perry, 1980) a water trap design was used in which the sticky plate was replaced by a bath of weak detergent (Figure 3.15). All traps in the same wheatfield site were mounted on telescope stands, and aligned with the wind direction. In this series traps were spaced in line along the wind direction as follows: (i) in pairs at spacings of 25–200 m; (ii) in lines of three traps at 50 m; and (iii) lines of five traps spaced at 25 or 100 m. The results showed that in (i) interaction was evident at spacings from 25–200 m, with the upwind trap usually catching more than the downwind trap; (ii) showed that the upwind trap caught more than the sum of the other two; and in (iii) the upwind and downwind traps caught considerably more moths than the other three, and this profile was similar at spacings of 25 and 100 m.

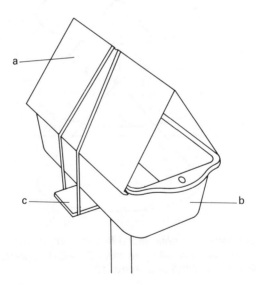

Figure 3.15 *Water trap design of pheromone-baited trap used in studies on pea moth (a, aluminum top; b, bin trap; c, stand) (Wall and Perry, 1980).*

These results indicate that the number of traps in a line had a greater effect on the profile of catches than the spacing between traps. They also suggest that the range of attraction to pea moths of traps baited with 100 µg of the sex attractant may be in excess of 400 m.

(b) Trap efficiency over short-term periods

In a further analysis of trap interaction under these conditions (Perry and Wall, 1984) each single afternoon trapping period — which formed the basis of the experiment above — was divided into a series of consecutive short-term intervals over that period, with trap collections being made at consecutive 5 min intervals for the three-trap lines, and 2 min for the two-trap lines. Triangular sticky traps were used in this experiment, a fresh plate being placed in each trap at the beginning of each interval, and replaced at the end of each period. The object of this experiment was to find out if, within each afternoon trapping period, differential changes in relative trap catch occurred, changes which might be related to larger changes in numbers of responding moths caused by shifts in micro-meteorological conditions.

These experiments confirmed that the mean catch over the whole trapping period conformed to the pattern defined in the earlier experiments, but the proportional catch in each trap varied widely within that period. While the proportion caught in the centre of the three-line array remained constant over the trapping period, the proportion caught in the upwind trap of two and three-line arrays showed trends in time. These trends differed between trapping periods, but followed a similar pattern in two lines of traps operated simultaneously. It is suggested that these short-term fluctuations reflect changes in that aspect of flight behaviour which is concerned with locating pheromone source, behaviour which in turn is affected, or determined, by a variety of subtle ecological, physiological and meteorological factors, all of which can change in a complex way in the course of the trapping period.

(c) Mark-release experiments

Further evidence of the range of attraction of pheromone-baited traps was obtained with the pea moth by means of mark-release-recapture experiments (Wall & Perry, 1987). Individually marked moths were released from a table placed at crop height, downwind of a triangular water trap. For each moth, time and direction of departure were recorded. As *Cydia nigricana* — unlike its close relation the codling moth, *Cydia pomonella* — is diurnal, its initial flight path could be observed visually until out of sight. These visual observations enabled changes in flight direction after release to be observed, including changes from crosswind flight to upwind flight. They also disclosed a greater flying height, on average 10 m but going up to 20 m, than previously observed. This behaviour may of course have been an artifact or escape reaction following confinement prior to release.

In the first experiment moths were released individually at 140–160 m downwind, and in the second experiment 500 m downwind of the trap which was baited with 100 μg of sex attractant. Moths approaching the trap were caught by handnet. Marked moths were recaptured at both these distances. Not only could they be recaptured at the maximum distance on the same afternoon, but there was also a high and sustained rate of attraction of males to the pheromone traps for 3–4 hours each day on several consecutive days. From this and previous evidence it was concluded that for this species the attraction of the pheromone trap over the open crop environment could extend as far as 1000 m.

3.3.6 Experiments on pea moth: trap design and pheromone plume

In each of the wheatfield experiments on pea moth described above, uniform trap designs were used, i.e. triangular sticky traps in the first series and water traps in the second, but an additional critical study on pea moth response to trap design has formed the subject of a separate investigation (Lewis and Macauley, 1976) which was motivated by the great need to understand the physical features which influence trap performance. Of the very wide range of traps in use by that time, about 20–30 had been used on tortricid moths alone — the group to which the pea moth belongs. Preliminary field trials had whittled down this number to the six most promising designs (Lewis *et al.*, 1975). These were as follows (Figure 3.16):

1. Pherocon IC. An example of a widely used commercially available trap for tortricids;
2. Triangular trap, which was tested when aligned both with the wind and across the wind;
3. Water trap;
4. Covered funnel trap;
5. Lantern trap designed to combine best features of the triangular trap and the water trap;
6. Open lantern trap; as above but modified to improve the air flow.

An additional feature tested was the effectiveness of various sticky retentive surfaces, comparison being made with five proprietary brands of adhesive using the triangular trap was standard.

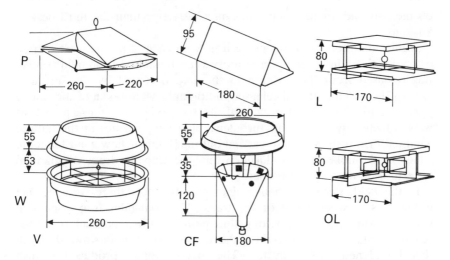

Figure 3.16 *Six designs of pheromone trap compared in wheat at crop level for pea moth. P. Pherocon IC; CF. Covered funnel; T. Triangular trap; L. Lantern; W. Water trap; OL. Open lantern (Lewis and Macauley, 1976).*

Judging the results on the basis of moths captured, the most successful design was the triangular trap and the least successful was the lantern design. When the number of moths caught was adjusted according to the total area of retentive, sticky surface in each trap, this resulted in a change in order of some traps, but the triangular, with an area of $180\,\text{cm}^2$, still remained superior to the others, including Pherocon IC — area $380\,\text{cm}^2$ — and the water model with an area of $532\,\text{cm}^2$. The triangular trap, when aligned with the prevailing wind, was the most successful.

In all traps the number of moths actually caught represented only a minority of those approaching. Many flew away before entering. Some entered to 'fan' their wings at the rubber stopper containing pheromone, and then flew off, while others managed to escape from the sticky surface after alighting on it. By counting the number of individuals arriving within about 2 cm of the trap, and comparing this with the number caught, it was found that in windy conditions only about one half of the moths approaching triangular traps entered them, and of those entering only about one half were caught, i.e. only about one in four arrivals at the trap were actually caught on first approach. The evidence is that many unknown features of traps deter moths from entering after they have been attracted.

Results of the sticky material tests showed that 'Bird tanglefoot' was far superior to other brands, being four times more effective than the second on the list, and about seven times more effective than the third best — Stickem.

Studies on pheromone plume pattern from each trap type were carried out both in the field and in laboratory wind tunnels. In the field, smoke was released from the position in the trap occupied by the sex lure. During the two minutes of release, photographs were taken of the smoke plume at 10 s intervals, for each of the six models. As these field tests were affected by natural air turbulence and irregularities in the natural wind pattern, tests were also carried out in a laminar flow wind tunnel in which cold smoke was produced within the trap by drops of titanium tetrachloride on a lint dispenser.

In the field, despite variables due to turbulence, there were distinct differences between some of the traps. The triangular trap orientated with the wind produced a long thin plume, with a particularly dense region extending one to two times the length of the trap downwind, which then broadened to a wispy tail. The covered funnel produced a much broader plume, while the water trap usually produced a diffuse cloud of smoke. The Pherocon produced an erratic plume, rapidly becoming diffuse even in partial cross wind. In the wind tunnel, with no air turbulence, the patterns were more clearly defined and consistent (Figure 3.17).

Films of moths approaching triangular traps showed a remarkable similarity between their tracks and the axes of plumes emitted from triangular traps both in turbulent and laminar winds. It appears that the numbers of moths caught by different traps were dependent on the shape of the attractant plume spreading from the trap.

3.4 Behavioural impact of multi-component pheromones

The study of pheromone trapping over the last 20 years falls roughly into two stages. In the first stage the main emphasis was on practical consider-ations of trap design and trap comparisons, and in assessing the role of pheromone trap catches in monitoring pest species of moth. In the second, more recent phase, the enquiry into specific insect response to pheromones and pheromone baited traps has probed much deeper, in line with the increasing need to find answers to many of the questions which arose as a result of the experience gained in the first phase. The fact that this more critical approach had followed after a long period of field ap-plication of pheromone trapping has resulted from the continuing increase in knowledge throughout that entire period regarding the chemical nature of the pheromones of different species. Perhaps even more important, is

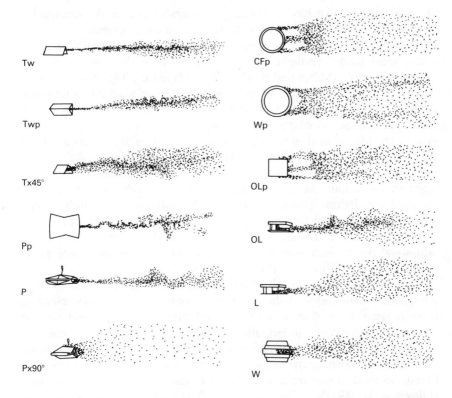

Figure 3.17 *Diagram of plume pattern traced from smoke plumes emitted from different trap designs in a wind tunnel (a, aluminium top; b, bin trap; c, stand) (Lewis and Macauley, 1976).*

the increasing awareness of the role played by each component of multi-component pheromones — which may contain up to six identifiable compounds — in the complex behaviour pattern of the male moth under the stimulus of pheromone produced by the female, or of synthetic pheromone emitted from a trap. This work has also highlighted the comparative rarity of those cases, such as the gypsy moth, in which the pheromone is a single chemical compound, also capable of being synthetised. (Cardé, 1984)

The situation by the mid-1980s has been summed up by Cardé and Baker (1984): 'Despite the rapid accumulation of identified pheromones and a description of the behaviour elicited, for a large number of pheromone systems we do not as yet know all of the chemical compounds involved, nor do we have a thorough understanding of the orientation mechanisms that are involved in many pheromone responses'.

In analysing insect response to pheromones and to pheromone-based traps, wind tunnel or flight tunnel experiments in the laboratory have played a key role (Miller and Roelofs, 1978) (Figure 3.18). A more recent design is simply constructed of rectangular pieces of glass and aluminium framing, 183 cm long and 61 cm wide and high (Glover *et al.* 1987). The wind speed in the tunnel is maintained at $0.25\,\mathrm{m\,s^{-1}}$. In the test, individual insects are placed in $3 \times 6\,\mathrm{cm}$ screened cylinders, with the open end of the cylinder upwind. After being allowed 30−60 minutes to take flight, a series of behaviour responses are recorded such as taking flight, orientated flight in the odour plume, initiation of upwind flight and, finally, attempted copulation. These are particularly instructive for moth species which, in addition to the main pheromone component, have other sex compounds that occur in small or trace amounts (Linn *et al.*, 1984). These additional compounds have been identified not only from female gland extracts but also from airborne collections.

In the cabbage looper moth, *Trichoplusia ni*, for example, four of these compounds were added to the previously known two-compound blend to form a six-compound mix which elicited male response equal to that shown to female glands, and significantly greater than to the two-component blend alone. An additional compound was also identified from female glands and airborne particles which, when added to the six-component blend, resulted in significant arrestment of the upwind flight of the male (Glover *et al.*, 1987).

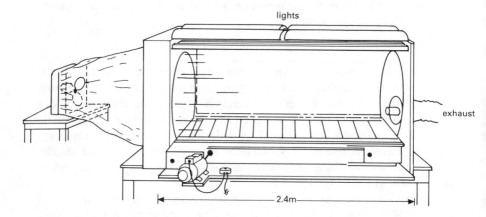

Figure 3.18 *Overall view of sustained-flight tunnel used in pheromone studies (Miller and Roelofs, 1978).*

The results of a series of subtraction assays with 5, 4, 3 and 2 component blends showed that whereas individual compounds exercised differential effect on male behaviour, the influence of any compound depended on the presence of certain other compounds, with several blend combinations eliciting peak response. This peak response was shown to the six-component blend, to five of the five-component blends, as well as to several four-component blends. The conclusion is that it is a blend of compounds, acting as a unit, which is critical in effecting optimal response in males.

These windtunnel experiments also show that in the interpretation of observations, account has to be taken of specific differences in the natural behaviour of males of different species. For example, the cabbage looper moth, the gypsy moth and the cotton bollworm moth fly up to the female to initiate courtship, while other species like the oriental fruit moth typically land and walk to the female.

These minor components play an important part in groups of species of the same family or sub-family which often utilize the same chemicals. For example, among the noctuid moths Plusiidae, the cabbage looper and the soybean looper share the same major chemical components. Both species produce two common minor components, but, in addition, there are further compounds specific to each species. When the role of these compounds was compared in a wind tunnel, as well as the role of common components in different proportions, it was demonstrated that discrimination of odour quality by male moths can result from minor components affecting male sensitivity, or arresting upwind flight response to the pheromone (Linn *et al.*, 1988).

3.5 Role of competition on trap efficiency

One of the factors influencing catches in pheromone traps which has received increasing attention is the competition for males between the synthetic pheromone in the trap and the natural pheromone exuded by the female population in the same locality (Knight and Croft, 1987). For a number of deciduous fruit pests it is often assumed that male response to pheromone traps coincides with female emergence and sexual activity. However, some workers have reported an absence of synchrony between catch, female emergence and sexual activity. In some studies it has been found that addition of virgin females to wild populations has produced reduced trap catch due to competition.

Experiments to clarify this problem were carried out on the moth

Argyrotaenia citrana (Orange Tortrix Moth) with the object of determining how such competition influences temporal trends in catch of males in pheromone traps, as well as in the attraction of males to live caged females. This experiment had to be considered against the background of established facts on the seasonal biology of this species, regarding male and female moth emergence over a single generation, as shown in a simplified model (Figure 3.19). Early in a generation, when male emergences exceed those of females, male density is much greater than that of 'calling' females. As a result, male captures in pheromone traps may be high because competition by females is low. Thereafter, during the peak of male and female emergences, male capturing by pheromone traps may be less efficient due to the preoccupation of males in locating and mating with the greater number of calling females. Later in the generation, when males cease but females continue to emerge — although in declining numbers — pheromone trap efficiency might increase again if females ceased calling after mating, and if males were relatively long-lived and continued to respond to the pheromone source.

Within a 50 × 25 m courtyard, with walls 20–25 m high, a single Pherocon (sticky) trap was set up at the beginning of each of four experiments. In each experiment 250 moths of each sex were released. In addition, females were housed in small screen cages within sticky traps.

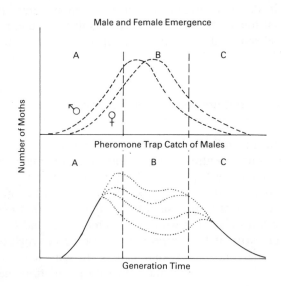

Figure 3.19 *Model of catch of* A. citrana *in pheromone traps in relation to emergence of male and female population (Knight and Croft, 1987).*

Over 50% of the males released were caught, 10% in the pheromone trap and 43% in female-baited traps. Over the first nine days, when male emergence exceeded female emergence, catches in female-baited traps, and mating frequency were low. In this period, the first catch in the pheromone-baited traps preceeded the first catch in virgin female traps by three days, reaching nearly its highest level for the whole experiment. During days 10−21, there was a peak of emergence of both sexes, and a peak of male catch in virgin female traps, and also a peak in mating activity. However, pheromone-baited trap catches remained constant, or dropped slightly, compared with the earlier period, i.e. the pheromone trap did not reflect increasing densities of male and female moths. The last 12 days were marked by a rapid decline in numbers of newly-emerged females, and particularly of newly-emerged males. There was a fall in catch in the female-baited traps, but a late seasonal increase in pheromone trap catches. Similar trends were observed in the following spring experiment, as well as in two summer generations.

These results show clearly that the initial peak in pheromone trap catch is usually not coincident with peak female emergence or mating. Instead, it considerably precedes these population events, and appears to be significantly influenced by competition with females, predominantly virgins. Allowing for various other factors such as variation in the synthetic pheromone blend, as well as the effect of temperature and other climatic components, it appears that these results reflect three distinct periods of pheromone trap performance in this species, dominated respectively by the earlier emergence of males, the cessation of calling by later females, and the ability of males to mate more than once.

3.6 Pheromones as disruptors to sexual communication

In unravelling the role of different chemical components of sex pheromones, the main emphasis has been on the part played by different components, or different blends of components, at each stage of male attraction with the object of increasing efficiency of pheromone-baited traps. In the course of that work it became clear that certain blends of pheromone, or certain concentrations, were not only non-attractive but actually inhibited male response at some point in the complex chain of events.

3.6.1 Specific response of closely-related species

This disruptive aspect of sex pheromones, or their components, plays a vital role in determining the specific responses of closely-related species of moth, which in many cases have a major pheromone component in common. In practical pest monitoring precise information about this disruptive role is required when the normal sex pheromone trap is found to be attractive to both the target species and to an allied species of no economic importance. This is exemplified by work on the winter moth, *Operophtera brumata*, an import into the USA and Canada from Europe (Pivnick *et al.*, 1988). The sex pheromone of this species is equally attractive to the two related indigenous species, the bruce spanworm *O. bruceata* and the western winter moth, *O. occidentalis*. Males of *brumata* and *bruceata* are difficult to distinguish, particularly in specimens damaged in sticky pheromone traps. In order to monitor the spread of the winter moth it is essential to make the pheromone trap more specific. The identification of the pheromone of the bruce spanworm in 1987 (Underhill, *et al.*, 1987) showed that it was the same compound as for the winter moth. The existence was also revealed of a compound which had an inhibitory effect − designated BSMI or bruce spanworm male inhibitor − which when added in amounts of 10−300 μg to 100 μg of pheromone, reduced the catch of bruce spanworm by an average of 83%, without interrupting the winter moth catch.

In order to improve the exclusion of native species, experiments were carried out in a wind tunnel; these showed that BSMI had a more potent inhibitory effect if males were in physical contact with the compound. In field trials in an area where only bruce spanworm occurred, pheromone-baited traps caught fewer *bruceata* when BSMI was placed on the outside of the entrance holes than when it was placed on the inside of the trap, the percentage reduction in catch being 97% and 82% respectively. In an area where the target species, *O. brumata*, predominated, BSMI, whether placed inside or outside the trap, did not affect captures of this species, pheromone-based traps without inhibitor being used as controls.

3.6.2 Pheromone release in small plots

Another aspect of the potentially disrupting role of sex pheromones is shown by the red-banded leafroller moth (*Argyrotaenia velutinana*) which has been the subject of research (Reissig *et al.*, 1978). This work showed among other things that in small field plots, pheromone released at the rate of $5 \, mg \, ha^{-1} \, hour^{-1}$ effectively eliminated orientation of this species to pheromone traps. In an extension of that work (Novak and Roelfs, 1985) the synthetic pheromone used as disruptant was dispensed from hollow capillary fibres in a treated plot. In both treated and untreated plots − $750 \, m^2$ in area − a Pherocon IC trap was placed in the centre and eight traps at the perimeter. There was no difference between treated and untreated plots in time at which activation and flight of males was initiated. However, the ability of activated males to orientate to pheromone-based traps was strongly affected. Regardless of whether males had previously been exposed to pheromone or not, significantly more moths were captured in the central trap in the control than in the treated plots.

3.6.3 Experiments in Australia on mating disruption

This disruptive role of pheromone chemicals is now receiving as much attention as their attractive role. In Australia, for example a considerable amount of research has been devoted to finding the ideal 'mating disruptant', i.e. a pheromone compound or blend of pheromones that would reduce the catch in pheromone traps, this being the first step in the search for a pheromone that will limit mating (Rothschild *et al.*, 1988a,b). The moths in question were the cotton bollworm *Heliothis armigera* (now *Helicoverpa armigera*) and the native budworm, *H. punctiger*, both key pests of cotton (Fitt and Van den Elst, 1988). One single and two multi-component blends of pheromone were evaluated in cotton plots in which pheromone was released from polyethylene dispensers placed at the rate of about 1000 per hectare. One particular compound, common to the pheromone of both species, reduced the catch of males of both species at traps baited with synthetic pheromone by 97% and 100% respectively.

Further trials were carried out on another species, the currant borer, *Synanthedon tipuliformis*, to find out the disruptive effect of various release rates (Rothschild *et al.*, 1988a,b). Blocks of 342 m^2 were treated with either 0, 16, 32 or 64 polyethylene dispensers tied in bundles of 0, 2, 4 and 8 respectively, so that the number of point sources from which pheromone was released remained constant. Each dispenser contained a 93:7 ratio of two pheromone components (an EZ acetate and a Z acetate). Each field block was monitored by five pheromone traps baited with the same pheromone blend. The trial that most effectively suppressed trap catch at synthetic baits was (E.Z)-2, 13−18 Ac) alone, a major constituent of the insect pheromone which reduced catch by 91%.

Rapid progress in the practical aspect of pheromone use is exemplified by the fact that a synthetic pheromone has been developed, and marketed as 'Isomate', for control of the codling moth by mating disruption (Bellas *et al.*, 1988). At the same time work still continues to determine the optimum pheromone blend and concentration for use in the essential monitoring trap used in these field trials.

3.6.4 Practical application of mating disruption with cotton pests

In other cotton growing areas such as Egypt, control of pests such as the pink bollworm, *Pectinophora gossypiella*, by mating disruption has now been practised for several years (Critchley *et al.*, 1983). In this case the mating disruptant is in microencapsulated form and is applied aerially; evaluation of these measures was done by means of pheromone-baited traps spaced at one trap per 10 hectares. Under those conditions moth catches could be as high as 1000 per trap/night at peak season. This blanket pheromone treatment was found to be effective in preventing male moths from being caught in pheromone traps, only when moth populations were low. From August onwards, disruption was no longer achieved.

In interesting contrast to the high moth catches recorded in pheromone-baited traps, the levels of population recorded in another cotton pest in that region, the cotton leafworm, *Spodoptera litoralis*, average only 5−9 moths per trap week; here a figure of over 10 per trap week was indicative of a large population increase (Ahmad, 1988).

There are now numerous field reports showing that disruption of orientation of male moths can be obtained by permeating the atmosphere with pheromone, or a blend or occasionally a single component. However, due to lack of knowledge of the disruptant mechanism involved, only a few compounds have been registered as chemical disruptants (Kehat *et al.*, 1982).

3.6.5 Mechanism of mating disruption in the red-backed cutworm

An example of one of the more recent investigations into these mechanisms is the work on the redbacked cutworm, *Euxoa ochrogaster*, in which the role of four components of the sex pheromone was tested in a wind tunnel and in field plot experiments (Paliniswamy and Underhill, 1988). In order to test the ability of males to locate target bait in the presence of disruptant dispensers, a control with target bait only was compared with the same target surrounded by dispensers. Using a series of concentrations of single components and their blends, male moths were released from cages positioned in the path of the chemical plume, and the flight patterns videotaped. In the field plots an additional method of observing night flight patterns was the use of a night-viewing device.

In the control experiment, in the absence of disruptant, the upwind flight of males, following the plume of pheromone which issued from the target septum, showed looping movements and zig-zag tacking. If they lost contact with the plume, males immediately engaged in searching flight, i.e. circular loops followed by even larger loops. In the presence of disruptant — at dosages of 1014 µg per septum — drastic behaviour changes occurred, none of the males initiating plume-orientated flight actually locating the target.

Chapter 4

Light Traps Versus Pheromone Traps

4.1 Introduction

In using trap capture data of any kind in order to monitor population changes of insect pests, or to determine flight and behaviour patterns, dependence on a single capture or sampling technique may provide information which is not only incomplete or biased, but which may also on occasions be quite misleading. The interpretation of capture data obtained by a single technique can be extremely difficult or speculative, and it has therefore been recognized that wherever possible the evaluation of a particular technique should be assisted by data obtained simultaneously by a comparison technique using quite different principles of capture or attraction. The interest in pheromones and pheromone-based traps has produced an immense amount of information about the performance of pheromone traps themselves, and their success or failure under a variety of conditions. In many cases the value of these figures can be greatly enhanced if information from other sampling sources is available.

Many of the moth species subject to intensive pheromone trapping trials were formerly monitored by means of light traps. The light traps themselves had proved valuable for some species, but less useful in other cases. In attempting any comparison of these two techniques for one and the same species, the potentialities and the limitations of each technique must be taken into account. The type of light trap used for moths normally depends on mains electricity, and this factor alone may limit the scope of this trapping technique. Light traps sample both sexes of moths, though not necessarily in equal proportions, while pheromone-baited traps capture the male component of population only. Some of these basic differences are well illustrated in the case of the spruce budworm (Miller and McDougall, 1973). In New Brunswick forest areas in Canada, light traps were formerly extensively used, their use being dictated by the fact that at that time there was simply no other technique for monitoring the adult moth population. In operations from 1960−64, light trapping in three of the five years produced no moths at all. In 1967 blacklight traps were introduced, and these were successful in detecting the increases in budworm moth density which led to subsequent high infestation and outbreaks. In these invasions 75−78% of the moths involved are females, which of course are undetected by the female sex traps, based on virgin females, which were in use at that time prior to the introduction of pheromone traps. Consequently, the latter failed to record the invasion.

4.2 Experiences in trapping the European cornborer (*Ostrinia nubialis*)

Several different invasions of the European cornborer moth in the USA have greatly added to the understanding of the relationship between these two capture techniques. In Iowa, where this species produced two moth flights per year, light trap data over 18 years had shown consistently that the second flight was the larger. In 1971−72 a study was designed, for both spring and summer flight, to find out whether European cornborer males attracted to the recently discovered synthetic sex pheromone (Z-11-tetra decenyl acetate) were synchronous with (i) males and females captured in light traps, and (ii) egg masses deposited in cornfields (Oloumi-Sadeghi *et al.*, 1975). Nine light traps, increased to 22 in 1972, were distributed in the experimental area, and daily collections were made from late May to early July, and again from mid-July to early August. Daily surveillance for egg masses in 20 cornfields provided an additional measure of population fluctuations. Two sticky types of pheromone trap were set up 100 m apart at the edge of each of the 20 fields, one trap per field being equipped with a filter paper cone and treated with cornborer sex attractant.

In order to minimize competition between different trapping systems, the light traps were set up at a minimum of 200 m from the sex pheromone traps. All female moths captured in the light traps were sex graded (Showers *et al.* 1974): I. Females unmated, ovaries gravid; II. Spermatophore full, ovaries gravid (captured within 24 h after mating); III. spermatophore partly depleted of sperm; and IV. spermatophore depleted of sperm, ovaries depleted of eggs.

The number of females within each class was compared with the number of egg masses on the 200 plants (10 per field), and with the number of males attracted to the sex pheromone traps. During spring flight, which lasted for 24 nights in the first year and 31 nights in the second year, in the nightly captures from light traps, class II females, i.e. those captured within 24 hours of mating, were selected for comparison. At light traps, the maximum capture of males was found to coincide with the maximum number of class II females, but few males were attracted to the sex traps within the period of heaviest mating and egg deposition. In the summer flight, during both years the majority of males captured in the light trap synchronized with the females captured, all classes combined. However, the synthetic sex pheromone did not attract appreciable numbers of males until after the disappearance of class II females from the light trap catch, and their replacement, through physiological shift, to classes III and IV.

Maximum egg deposition occurred within 1−3 days of the maximum of class II females in the light trap. In the pheromone trap, the male maxima which synchronized with the maxima of all females classes combined, occurred after the majority of egg masses had been deposited in the field. If the pheromone traps had been competitive with the number of feral females available for mating, a capture response simulating that of the light trap would be expected, but the only period at which captures of males in pheromone traps approached synchrony with captures of males and females in light traps was when the moth population was low, and few feral females were available for mating. Very similar conclusions have been reached in the case of other species such as *Heliothis zea* and *H. virescens* (Roach, 1975).

In other studies on pheromone trapping of the European cornborer the general pattern which emerged was that pheromone traps attracted significant numbers of males in the first, spring, flight, but very few in the second flight (Roelofs *et al.*, 1972; Cardé *et al.*, 1975). In a comparison made over two seasons in western Massachusetts with two blends of pheromone versus blacklight traps, both traps reflected the beginning of the first flight in mid-June, at which period the pheromone trap caught considerably more than the light trap (Fletcher-Howell *et al.*, 1983) (Figure 4.1). From the last week of June until mid-July, the light trap recorded some activity which was not replicated in the pheromone trap.

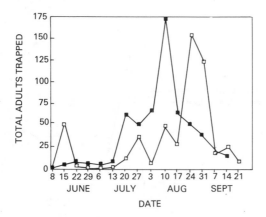

Figure 4.1 *Pheromone catch □ and light trap catch ■ of European cornborer moths (Fletcher-Howell et al., 1983).*

In the second flight, the peak pheromone trap capture was on 24th August, 2 weeks after the light trap peak, possibly due to depression of pheromone attraction caused by competition with feral females.

Studies on the European cornborer in North Carolina also indicated that pheromone-baited traps perform poorly in relation to blacklight traps (Figure 4.2) and that catches in pheromone traps peaked after the light trap maxima (Kennedy and Anderson, 1980). The reduced efficiency of the pheromone traps was attributed to competition for males by feral females.

4.3 Comparison involving two different types of pheromone trap

In a later study (Thompson *et al.*, 1987) two different types of pheromone trap were compared with blacklight trap captures. One of the pheromone traps was the conventional sticky type, while the other was an aerial water-pan trap which contained equal parts of water and anti-freeze, and a small quantity of dish soap. Results (Figures 4.3 and 4.4) show that in the comparison between sticky and blacklight, both traps reflected changes in population density, although the levels of capture were very different, 4.3 males per week compared with 473 for the light trap. In the comparison between aerial water trap and blacklight it was shown that at low population levels catches by these two methods were not significantly different, compared with the significantly lower levels in the sticky sex trap. At high

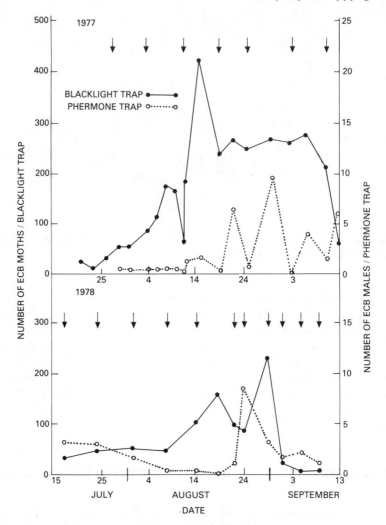

Figure 4.2 *Number of European cornborer moths (males and females) caught in blacklight trap compared with catch in pheromone-baited trap (Kennedy and Anderson, 1980).*

densities, this correlation is lost, increasing populations as measured by blacklight not being matched by increases in pheromone trap catch. Night observations on the two types of pheromone trap revealed that the aerial water trap was 400% more efficient at capturing European cornborer

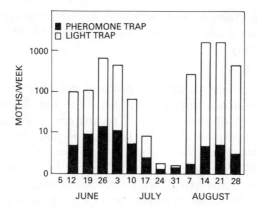

Figure 4.3 *Comparison of weekly catches of European cornborer moth in conventional sticky-type of pheromone trap and in blacklight trap (Thompson et al., 1987).*

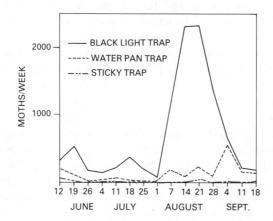

Figure 4.4 *Comparison of weekly catches of European cornborer adults in a high population area by (i) blacklight trap, (ii) water-pan pheromone trap, and (iii) sticky pheromone trap (Thompson et al., 1987).*

than the sticky sex trap. Moths were observed to fly into and out of both types of trap, but in the case of the sticky design several moths that touched the sticky surface managed to escape, leaving only a few scales or a leg. In the aerial water pan trap, moths which touched the liquid did not escape, and in fact were killed in seconds.

4.4 Seasonal changes in reaction to the two trap types. Experiments with the w-marked cutworm (*Spaelotis clandestina*)

This species presents an unusual problem not only in its differential reaction to pheromone traps and to light traps, but also in the fact that reactions to each of these trapping techniques is subject to seasonal variation. The background to this situation is that these moths, found abundantly throughout North America, emerge in June, but after an active period of flight go into local hiding for 4–8 weeks. During that period they apparently experience a retarded development. In the latter part of August the moths re-emerge and fly freely until mid-September. During both periods of active flight these moths could be taken regularly in blacklight traps. However, Pherocon ICP traps baited with blends of synthetic pheromone and exposed throughout that period from June to September, were only attractive to males from mid-August onwards, with none captured during June and July. From mid-August until the end of flight in mid-September, attraction of moths to pheromone traps was stronger than to light traps (Steck *et al.*, 1982).

Females taken in June in light traps adapted for the retention of live captured moths, and caged in pheromone traps, attracted no males to these traps in June and July, even though moths continued to be attracted to the light trap during this test period. In contrast, females from June light traps, kept over summer, attracted appreciable numbers of males in the 3rd week of August. It appears from this that *Spaelotis* probably do not emit a main pheromone during the pre-aestivation period. Nor do males respond to the chemical, or to females, during that period. The *S. clandestina* sex pheromone thus appears to have a single component which is seasonally dependent, and in this respect atypical of noctuids in general. Traps baited with the synthetic pheromone can be used successfully only in the post-aestivation or reproductive period; they fail to attract males in the pre-aestivation period. Another aestivating noctuid moth, the army cutworm, *Euxoa auxiliaris*, also fails to respond to synthetic pheromone during the pre-aestivation period (Struble and Swailes, 1975; Struble *et al.*, 1977).

As a biproduct of this comparison, it was found that light trap captures, operative throughout the entire season, were always found to be higher in the second flight period than the first, even though the same generation is involved in both flights. This is explained by the fact that the daily dark period during which attraction to light can occur, is about 10–11 hours at the end of August, compared with 5–6 hours in the latter part of June.

This incidentally providing a very instructive example of seasonal changes in effectiveness of light traps.

4.5 Comparison of blacklight trap with two types of pheromone trap for the western bean cutworm (*Loxagrotis albicosta*)

Blacklight traps have long been used to monitor this pest species in the USA, although their limitations were recognized. Variations in damage caused by this pest from field to field reduced the value of the light trap in predicting what would happen to individual fields. Other limitations are the dependence on a power supply, and the fact that light traps can attract large numbers of non-target insects. The isolation and identification of the sex pheromone compounds of western bean cutworm as late as 1981 dramatically changed this situation (Mahrt *et al.*, 1987).

A comparison of blacklight performance was made with two different pheromone traps. The first was a standard Japanese beetle, water-filled trap with a few drops of detergent to break the surface tension and enable moths to be drowned. The second was a water bucket, also with water and detergent. Dichlorvos was used as the killing agent in the blacklight trap. The results (Figure 4.5) show that in the peak flight months, roughly from mid-July to mid-August, pheromone traps initially lag behind light traps, reflecting a predominance of females. After this, when males predominate, pheromone trap catches overtook the light trap catches. A further comparison of both trapping systems was made at different heights, 3.7, 2.4, 1.2 and 0.6 m, above ground. This showed that for both traps, catches at the two lower levels were significantly higher than at the two higher levels. The most appropriate height for monitoring was 1.2 m.

4.6 Observations on *Helicoverpa armigera* in India

The comparison of these two techniques has been the subject of a particularly instructive study in India (Dent and Pawar, 1988). It arose from the recognition that nocturnal activity of some species of Lepidoptera occurs in a definite sequence during the night (Lingren *et al.*, 1978; 1980). This may take the form of an initial period of oviposition interspersed with feeding in the early part of the night, followed by increase in mating activity as the night progresses. From this it follows that the modes of activity and the sequence in which they occur could affect the relative

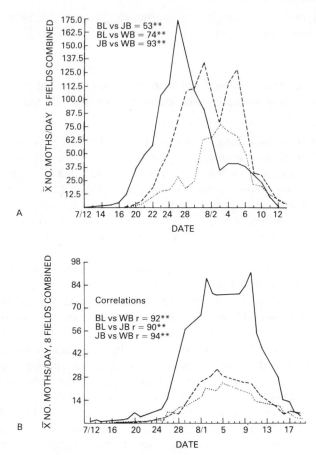

Figure 4.5 *Comparison of catches over 3 years of western bean cutworm by blacklight trap and two different pheromone traps, viz water-filled trap (standard Japanese beetle type) and water bucket type (Mahrt et al., 1987). A. 1982; B, 1983. Blacklight —— Water bucket – – – Japanese beetle type.........*

likelihood of capture in a trap at any particular time during the night. The timing and size of pheromone trap catches during the night may reflect the timing and extent of the natural activity of males. This, and other modes of activity, also influence the likelihood of capture in light traps.

In the comparison of the two main capture techniques a modified Robinson light trap was matched against two funnel pheromone traps, recording hourly trap collections from 19.00 hours to 06.00 hours for each night of the lunar cycle throughout a complete month. The two pheromone traps were placed 150 m on either side of the light trap, at a height of 2 m. The results (Figure 4.6) showed that in a total light trap catch of 1892 males and 2305 females, there were two peaks in the month which occurred during the period of no illumination in the lunar cycle. In contrast, in the pheromone traps, in which the more efficient of the two captured 1172 males, numbers fluctuated every few days but showed no distinct peak.

During the cycle of night activity, the two trapping techniques present a different picture. In the light trap the mean number of adults caught per hour increased steadily during the night, with peak catches of both males and females at 03.00 hours (Figure 4.7). In the pheromone trap the mean hourly catches were more evenly distributed throughout the night. A temperature of 12°C appears to be a critical zone for both light and pheromone captures, catches being little affected above that point, but

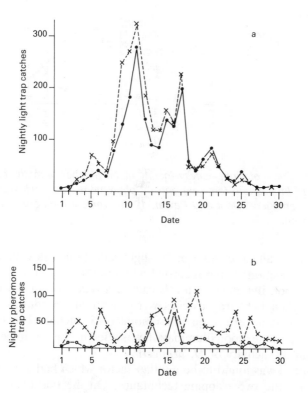

Figure 4.6 *Comparison of captures of* Helicoverpa armigera *by two main capture techniques, (a) Robinson light trap (males and females), and (b) two funnel pheromone traps (males only) (Dent and Pawar, 1988).*

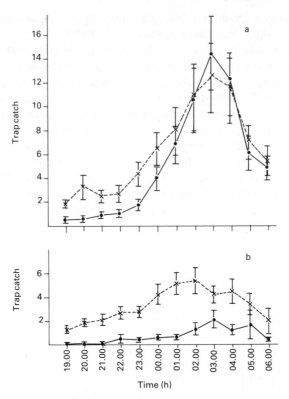

Figure 4.7 *Cycle of night activity (hourly mean number of moths caught) by (a) light trap (males and females) and by pheromone trap (males only) (b) of* Helicoverpa armigera *(Dent and Pawar, 1988). Vertical bars are standard error of mean catch.*

rather dramatically affected even by slight reductions below that point, no light trap collections being recorded when the temperature fell to 11°C.

Higher moon illumination levels caused a reduction in the number of occasions when light trap catches were recorded. In contrast, moon illumination had no effect on pheromone trap catches. The implication from this is that the moon illumination was affecting light trap performance rather than the activity of *H. armigera*.

Wind speed was found to be another factor which had a discriminating influence on the two trapping techniques. On the one hand, light trap catches were favoured at low wind speeds, and catches of both males and females declined quickly at wind speeds above 4 km/h. On the other hand, pheromone trap catches were consistently high when wind speeds ranged between 5 and 10 km/h. This is almost certainly due to the fact that the size of the pheromone plume — the active area of the trap — is dependent on wind speed; the higher the wind speed, the more likelihood of insects flying into the plume, and hence being attracted to the trap.

4.7 Australian experiences with trap comparisons

Certain limitations in the use of pheromone traps for monitoring populations of the cotton pests *Heliothis* spp. in Australia were also revealed when capture data were compared with that of a light trap (Fitt and van den Elst, 1988). On the important practical aspect of short-term prediction of oviposition activity within the crop, light traps were found to provide a reasonably accurate indication both of egg laying and of adult populations. But in dry funnel designs of pheromone trap, the species composition of *Heliothis* does not correlate closely with the species composition of eggs laid in adjacent cotton fields.

Further insight into pheromone trap performance was made by means of direct observation, using night vision goggles (Fitt *et al.*, 1989). This disclosed that under these Australian conditions at least, pheromone traps proved surprisingly inefficient in catching males attracted to them. Males appeared to be actually deterred by the presence of the trap itself, and would approach much closer to lures exposed without a trap. Catch efficiency also varied with moth density; at high densities males attempt to drive one another way from the pheromone source as they do in the vicinity of 'calling' females.

Chapter 5

Flight Traps and Interceptor Traps

5.1   Introduction

Of the wide range of techniques available for capturing and sampling insects in flight are methods which are not dependent on attractants such as light, odour or visual stimuli. The use of passive, non-attractant techniques is essential when trying to determine natural flight paths and flight patterns of insects in search of their plant or animal hosts, or other sources of food, as well as flights related to mating patterns and breeding foci. As these natural flight patterns can be easily disrupted by attractant traps, or even by suction traps, the ideal would be a completely passive, invisible, inert interception design of trap. Several traps have been designed with this clear objective, forming a vital constituent of the sampling armament.

The evaluation of these techniques has proved rather more complex than visualized. This is due to two main factors. First, some flight traps originally believed to be non-attractant can, under certain conditions, exert a visual effect on both day and night-flying insects. In some cases this effect has proved to be attractant, while in others repellent. The other factor is the increasing tendency in trapping trials, not closely concerned with the strict definition of flight patterns, to incorporate additional capture methods in the 'non-attractant' flight interceptor, either in the form of suction traps, CO_2 or both. Such methods, designed to boost the catch in the flight interceptor, inevitably introduce new factors which make direct comparison between 'attractant' and 'non-attractant' trap data rather less straightforward.

5.2 Malaise traps

The Malaise trap, designed nearly 30 years ago (Townes, 1962) has proved to be one of the most useful and widely-used non-attractant, static insect traps, and one that is equally operable by day and by night. The basic design (Figure 5.1A), which has changed very little over the years, is an arrangement of converging screens or baffles which mechanically intercept the flight of insects, and channel the trapped insects into an apical non-return collecting bottle or container. Because of its ease in packing, assembling and dismantling, as well as its low cost, it has been particularly favoured in insect faunal surveys under a wide variety of conditions. Despite its wide use however, very little work has been done in evaluating its collecting efficiency (Darling and Packer, 1988). What has been the subject of considerable experimentation and field trials has been the efficiency of Malaise traps when supplemented by additional trapping devices, such as the use of attractants suitable for the larger blood-sucking flies, e.g. horse flies (*Tabanus*) and deer flies (*Chrysops*). The basic design has also been enhanced by the use of pans of water at ground level in order to retain those insects which, in the absence of an attractant, might otherwise escape from the trap.

These supplementary methods have to be taken into account in making any precise comparison between this trapping technique and other 'non-attractant' flight interceptors. They have also to be considered when making comparisons between the Malaise trap and the visual or silhouette type of trap as exemplified by the Manitoba trap (section 7.4), both of which are used for sampling the same group of strong flying biting flies.

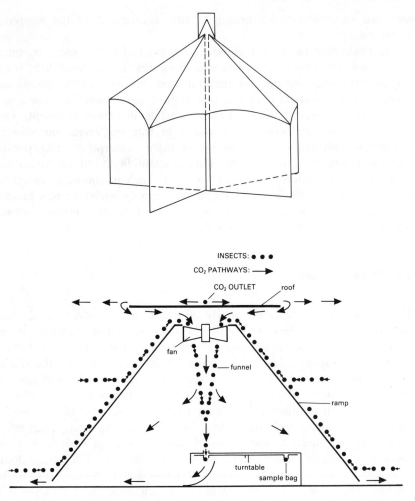

Figure 5.1 *A. Malaise, trap, original design. B. Pyramidal centre of composite Malaise type trap, provided with ramp and extended baffles, showing lines of attraction of CO₂ bait (Trueman and McIver, 1981).*

Critical studies on the effectiveness of Malaise traps have been mainly carried out on two different groups of insects, the horse flies and deer flies noted above, and a range of wasps and allied Hymenoptera.

5.2.1 Experiments with horse flies and deer flies

Horse flies and deer flies are well known biting pests of both man and domestic stock in many parts of the world, but it is mainly in the USA and Canada that the reactions of this group of flies to the Malaise trap have been critically examined, under a variety of conditions (Anderson *et al.*, 1974; Roberts, 1972, 1975, 1977). A great deal of ingenuity has gone into improving design and performance not only for special requirements, but also physical damage and wear-and-tear to which traps are likely to be exposed under different climatic conditions. A useful starting point is a simple comparison between the Malaise trap, which captures both sexes, and aerial netting or sweeping for the collection of female biting flies attracted to bait, usually the human collector himself (Tallamy *et al.*, 1976). In order to establish some uniformity in capture by aerial sweep net, samples were taken regularly along fixed routes by making ten overhead sweeps with the net, repeated three times, every 15 min. Equal catching time was spent in each of five study areas, and covered the entire daylight period up to and beyond dusk. One Malaise trap was placed in a tabanid flight path at each of the five sites.

The results showed that aerial netting sampled all dominant deer flies (*Chrysops*) far in excess of those taken in Malaise traps. For example, in one season a total of 7594 *Chrysops vittatus* were net-caught, compared with only 96 trapped in the Malaise. In contrast, the Malaise trap sampled greater numbers of all dominant species of *Tabanus*, horse fly, with one exception.

In another study in Mississippi (Roberts, 1976b) the comparative effect of six trap types, including two types of Malaise trap and two designs of canopy trap (Adkins *et al.*, 1972) were tested; five of the six traps were tested in pairs, one with additional attractant CO_2. The results showed that different species tend to react to different trap designs in different ways, but they were unable to indicate clear-cut reactions which could assist in determining behaviour patterns.

There are clearly many factors determining the efficiency and performance of Malaise traps; among these, colour and age of trap appear to be important (Roberts, 1972, 1975). Quite small differences or changes in the traps are also important in determining how effectively the flies directed into this interceptor trap are guided into the apical collector, or whether they manage to fly back out the way they entered. The Malaise trap on its own has been used to study vertical flight patterns of tabanids (Roberts 1976a), the trap being modified by reducing the size of the opening from the normal 1.2 m high to 0.6 m, then placing the trap on a screened platform which prevents tabanids from entering from below. Traps were set up or suspended at five levels, from 0 to 0.6 m near ground level, to 3.6−4.2 m. The results showed that, depending on the species, 72−88% of the flies were trapped at the two lowest levels. Of these low levels, the majority were taken between 0.9 and 1.5 m. The results gave a clearer picture of the concentration of tabanids in a flyway, where only a low percentage appear to fly above 1.8 m. There was also little tendency for flies to be attracted above that level even when positive attraction was provided by supplementing the Malaise trap with CO_2.

More recently the use of Malaise traps on an extensive scale has played a vital part in studies on the flight range and dispersal of a particular tabanid, *Tabanus abactor*, in Oklahoma (Cooksey and Wright, 1987). For this purpose 12 traps were placed in a circular pattern at distances of 0.4 km (eight traps) and 0.8 km (four traps). Traps were baited with dry ice which allowed the CO_2 to escape from under the trap. In this study it became evident that trap location was an important element in determining capture rate, the extra attraction of certain traps being apparently related to the fact that in its flight paths this species shows a marked preference for the forest edge.

5.2.2 Modified Malaise trap for *Simulium*

A type of trap embodying many features of the Malaise trap has been designed for a wide range of biting flies in Canada (Trueman and McIver, 1981). For the purposes of that investigation it was necessary to design a trap which would segregate catches according to small time intervals. Rather like the Malaise trap, this takes the form of a pyramid, 4 ft high and 8 ft wide, with radiating baffles extending out 8 ft from the base of the pyramid, which itself incorporates multidirectional ramps. The baffles increase the effective trapping area from 64 to 400 sq.ft. Capture efficiency is achieved by fitting a 12 in Vent-Axia suction trap in the apex of the pyramid. Inside, the fan is connected to a fine-mesh tapering funnel leading to the segregating mechanism. As soon as each collecting bag moves on the turntable box, out of the air stream, it is exposed to a high concentration of dichlorvos which ensures that collected insects are immobilized and not subject to damage. The supplementary attraction for the biting flies is provided by CO_2 released at the rate of 700 ml/min. This trap captured 20 species of mosquito and 29 species of tabanid horse flies, as well as over one hundred other insect families, including aphids, cicadellids and cecidomyids. The effect of the added CO_2 and the incorporation of a fan into a conventional Malaise type trap is illustrated in Figure 5.1B, which shows the lines of attraction to mosquitoes.

5.2.3 Design of trap used for Hymenoptera

In the studies on the reactions of various Hymenoptera to Malaise traps in Ontario, Canada (Darling and Packer, 1988) yellow pan traps were recessed into ground level along the long axis of the trap, and filled with water and detergent. This allowed a comparison to be made between insects caught in the conventional collecting bottle at the apex of the trap, and collections from the pans themselves. There were therefore two quite distinct factors determining the reactions of insects approaching the trap; first, random flight into the conventional open section of the Malaise trap, and secondly, visual attraction to the yellow pans. In addition, there is probably a certain random element involved in the pan catch of trapped insects trying to escape, before their flight can take them upwards into the apex of the pyramid.

In trap comparisons, in which two traps operated simultaneously, additional alternatives were provided in that while the aperture of the mesh netting was the same, i.e. a maximum opening of 0.8 mm, one trap had fine fibres of a uniform drab green while the other had coarse fabric of a different weave, and had its lower panels black and the upper ones white.

The Hymenoptera collected were sorted into sawflies, parasitic wasps and stinging wasps. The parasitic wasps were further subdivided into the larger Ichneumonidae and Braconidae, and the smaller microhymenoptera. These different groups were found to react in different ways to the two trap components. For example, the stinging wasps, Sphecidae, were equally distributed between pans and trap heads, while Pompilidae were better represented in the pans. In the microhymenoptera, about equal numbers were taken in each element. It was found that regardless of trap location the coarser bicoloured trap recorded high catches with the aculeates, while the fine mesh traps were more effective with microhymenoptera. To what extent these reactions were influenced by differences in air flow through the two mesh systems, or by visual response to the two shades concerned, was not determined.

5.3 Flight traps and interceptor traps for mosquitoes

One of the first problems to be tackled by the long-term mosquito research project in the Gambia, West Africa (section 2.4.2) was to devise a non-attractant type of flight or interceptor trap which could be used to define the flight paths taken by hungry mosquitoes between their breeding-grounds and their blood meal source. This applied in particular to defining this flight path in relation to distance from the warm-blooded hosts (Gillies and Wilkes, 1970, 1972). As the question of direction of flight, whether in relation to the host or to wind direction, was of primary interest, it followed that a passive sampling technique had to be used which could effectively capture and retain mosquitoes entering the trapping

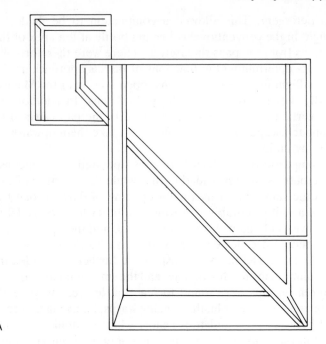

A

Figure 5.2 *Ramp trap design as originally used in mosquito flight studies in the Gambia (Gillies, 1969) (A) and 2-tier design (B).*

area from specified directions. This aspect of trapping is complementary to the studies on suction traps later carried out by the same team, and described in section 2.4.2.

5.3.1 Design of ramp trap

The ramp trap which was first devised for this purpose (Gillies, 1969) took the form shown in Figure 5.2. Mosquitoes entering the wide opening of the trap tend to fly upwards on the incline of the ramp, and their flight takes them into a cage through a horizontal slit opening which can be opened or closed as required. Later an improved and modified form of ramp trap was used (Figure 5.3) in which the ramp was separated from the cage unit (Gillies and Wilkes, 1970). The netting funnel of the ramp section was now extended outwards so as to increase the size of the entrance to approximately 6 ft high and nearly 6 ft wide, thus providing a greater catching area. Traps were operated overnight, with the entrance closed by day when not in use.

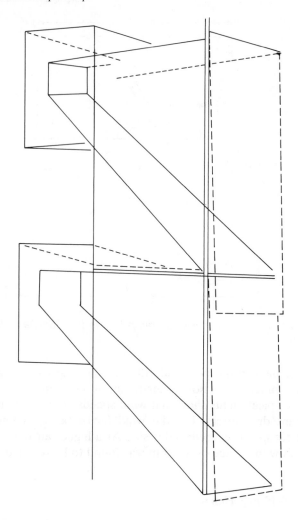

B

Figure 5.2 *Continued.*

Although these traps planned to be used in a wide range of experiments were designated 'non-attractant', the operators were fully aware of possible factors which might influence their performance. At this stage the possibilities of visual effects in moonlight were recognized, but it was not until later that possible visual impact of these traps was critically tested, both on moonlit and on moonless nights. At this initial stage, rather more attention was given to another physical factor which might influence trap

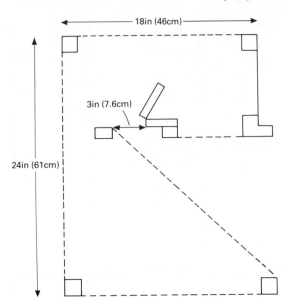

Figure 5.3 *Later improved design of ramp trap in the Gambia (Gillies and Wilkes, 1972).*

performance, namely the potential effect of the trap itself in altering or reducing the immediate airflow or airstream in the vicinity of traps in the field. Many tropical nights show that wind speeds are too low to produce any significant drag effect, but the wind factor becomes important at measurable air speeds exceeding 1 ft s^{-1}. At a higher air flow of 2–5.5 ft s^{-1}, the air flow in the lee of the trap was found to be reduced more than 50%.

5.3.2 First field experiments

Various systems of ramp trapping were set up in open areas at a location 50 miles inland from the mouth of the Gambia River, where the water is still brackish and the river lined with extensive mangrove swamps. The great numbers of mosquitoes produced from these swamps invade the villages which are separated from the swamps by a distance of 1–2½ miles. The traps, which were used either in the absence of bait or at various distances from available bait, offered unimpeded entry to mosquitoes flying towards the villages. The layout of the first trapping system (Figure 5.4) involved avenues cleared in the bush in the form of a cross. The attractant baits provided were penned either at the outer end of one

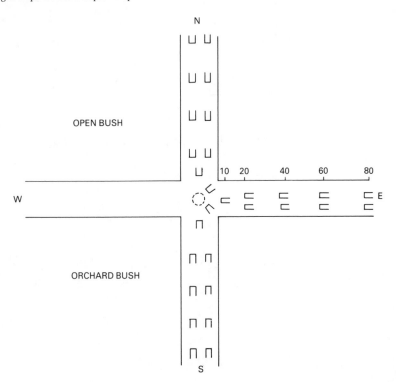

Figure 5.4 *Layout of trapping system using ramp traps as used in the Gambia. The interrupted circle represents the enclosure of the baits, and the numbers indicate the distance from the baits of successive lines of traps (Gillies and Wilkes, 1969).*

of the avenues, or at the centre of the cross. In each ramp or interceptor traps were set at increasing distances outwards from the bait up to 80 yds.

Prominent among the rich mosquito fauna of this area was *Anopheles melas*, the brackish-water member of the *Anopheles gambiae* complex and principal vector of malaria in Africa. From the total number of mosquitoes trapped at various distances from the bait, it was possible to construct 'catch curves' according to the two alternative baits used, animal bait and CO_2. Analysis of catches showed three types of reaction (Figure 5.6) according to the extent to which the high catches near the host attractant fell off with increasing distance, eventually reaching a constant level.

In the following year a more extensive trapping system was designed (Gillies and Wilkes, 1970) with regularly spaced traps arranged in a radial

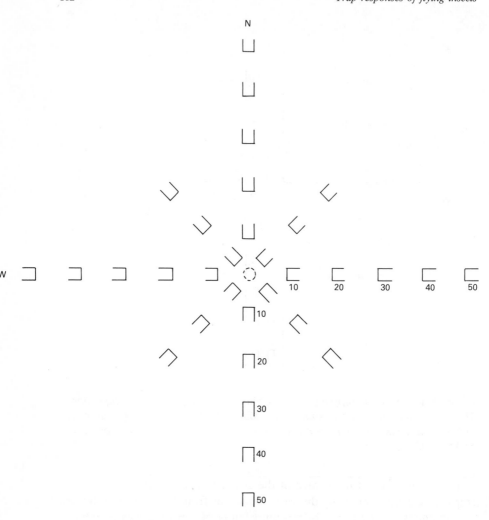

Figure 5.5 *Later trapping system on a radial basis (Gillies and Wilkes, 1970).*

Figure 5.6 *(Opposite). Reactions of four different species of mosquito with regard to relation between host attraction and increasing distance from host, as shown by 'catch curves'. (Gillies and Wilkes, 1972).*

pattern (Figure 5.5). This design was then adopted for intercepting mosquitoes flying within 4.5 ft of the ground, and approaching from any point of the compass. In a further stage in the project, experiments were extended inland to a point 200 miles from the mouth of the river, into a

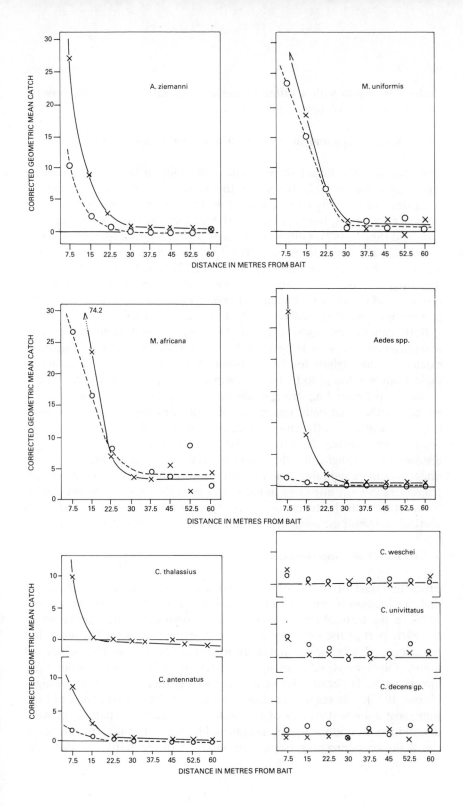

freshwater region with a different mosquito fauna, which again showed three main types of reaction or 'catch curve'.

5.3.3 Ramp traps and the vertical distribution of mosquitoes

The second main aspect of mosquito behaviour to be inevestigated by means of ramp or flight traps was the vertical distribution of flight. A great deal was already known about this aspect of mosquito behaviour in tropical rain forests, but very little work had been done in open savanna or farmland (Snow, 1975; Gillies and Wilkes, 1976). Flight traps were stacked at three levels, from the ground to 1.37 m (i.e. the level used in the bait attractant tests described above), 1.45 to 2.82 m, and 2.90 to 4.27 m. At this stage in the investigation, suction traps were added in order to provide additional non-attactant sampling methods (section 2.4.2). The two methods were run concurrently at the farm sites.

Both capture methods revealed the same vertical flight pattern, in *Anopheles melas*, with largest catches at the lowest levels and smallest catches at the highest levels. However, the total number taken in the flight traps was found to be trivial compared with the large catches in the suction traps, providing strong evidence that in the absence of any attract-ant bait in the flight path, this species was showing avoidance of the flight traps. In contrast, in the case of *Culex thalassius* and *Culex decens*, while suction trap catches appeared to be uninfluenced by the presence or absence of moonlight, the flight trap catch recorded much higher catches under moonlit conditions. The indications were that these 'non-attractant' traps were visually attractive to these species in the presence of moonlight, by contrast with the reactions of *Anopheles melas*. In further studies on vertical distribution, suction traps alone were used (section 2.4.2a).

5.3.4 Effect of wind direction

From the start of the Gambia investigations the question of mosquito flight in relation to wind direction received prior attention (Gillies *et al.*, 1978) In the tropical environment of the experimental site, conditions in the early part of the night are ideal for studying this factor because wind direction and wind speed normally maintain constant levels for several hours. Flight traps were set up in a circle of 4.5 m radius, facing eight directions, and operated during a $3\frac{1}{2}$h period in the early part of the night (Snow, 1976). The results are shown in Figure 5.7 of unfed females of the four most abundant species of mosquito. This charts the pattern of wind distribution around the mean, as matched by the distribution of catches in the flight or interceptor traps. The charts indicate fairly precisely the

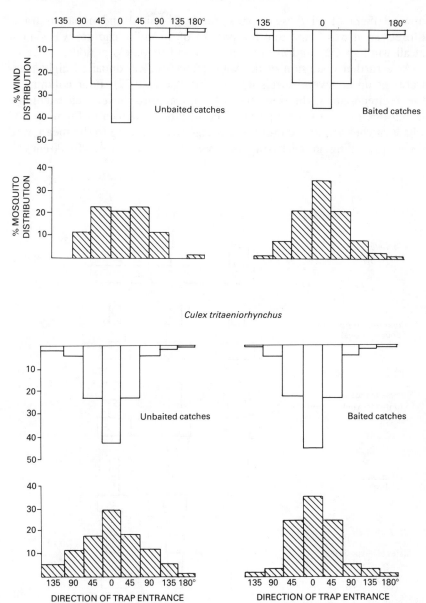

Figure 5.7 *Distribution of mosquitoes round a circle of eight directional suction traps ▨, and the opposing wind direction. Numbers of unfed females taken at baited and at unbaited traps. (i) Anopheles melas; (ii) Culex tritaeniorhynchus (Snow, 1976).*

upwind flight of all these species, for both baited and unbaited traps. Moreover, it was found that the pattern of mosquito capture was similar at all wind speeds, and at both moonlit and moonless periods.

As a further extension of the wind effect studies, unbaited flight traps were set up at several levels up to 14 ft (Snow, 1977). Four columns of three directional flight traps were set up at three levels, each column facing one of the cardinal points. The results are shown in Figure 5.8 in which catches are sorted into three categories according to the mean wind speed prevailing at each trapping period, $0-40\,\text{cm s}^{-1}$, $40-80\,\text{cm s}^{-1}$,

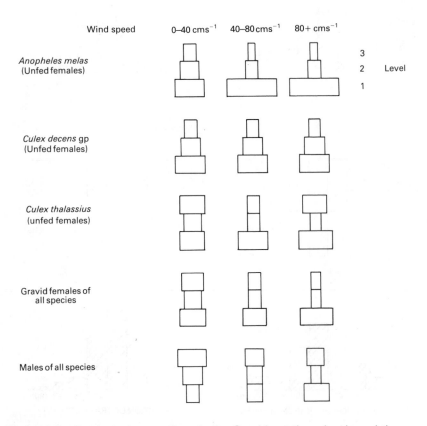

Figure 5.8 *Catches of mosquitoes in the Gambia at three levels and three wind speeds. Combined results of two experiments with three tiers of flight traps, and one experiment with non-directional suction trap (Snow, 1977). Level 1. Flight trap at 0.00 to 1.37 m. Suction trap at 0.68 m. Level 2. Flight trap at 1.45 to 2.82 m. Suction trap at 2.13 m. Level 3. Flight trap at 2.90 to 4.27 m Suction trap at 3.51 m.*

and 80 + cm s^{-1}. In the case of *Anopheles melas*, figures for flight in relation to wind had to be based mainly on suction trap data as in the absence of bait this species tends to avoid the passive, but visible, flight trap. At this stage in the investigations a new sampling device in the form of the 'directional suction trap' was introduced, results of which are discussed on page 82).

For the second objective of this series, direction of flight at different heights according to wind direction, direction traps were again used at three levels, and facing four directions. In one series, one of the four traps faced directly upwind, and one downwind, while two faced directly across wind at 90°. In the second series, the wind was diagonal to the axes of all the traps, with the two upwind traps being placed at 45° to the wind. The combination of these two arrangements is virtually a replica of the original flight direction test at ground level using a circle of eight flight traps (page 164) supplemented in this case by additional data from two different height levels.

The results (Figure 5.9) show a general pattern of upwind flight at all three levels on the part of unfed females of the three most abundant mosquito species. The largest catches were taken in the traps facing directly downwind, with the distribution around this maximum inversely reflecting variations in the wind direction. The reactions of *Anopheles melas* however showed some differences in that flight appeared to be predominantly upwind near the ground, but with an increasing tendency towards downwind flight at higher levels.

5.4 Window traps for bark beetles

Simple interruption of flight path in the absence of any visual element has long been practised in tne case of beetles. These take the form of window traps or glass barrier traps, and many groups of beetles have been caught simply by means of a single pane of glass above a trough. In some studies, house windows 38 cm high and 72–97 cm wide, provided with galvanized rain gutters, have been used successfully to capture a wide range of insect fauna in fir forests (Canaday, 1987). In a further development of the single pane, four glass panes have been attached to a central axis, producing a baffle or vane trap (Hines and Heikkenen, 1977)

5.4.1 Experiments with transparent acrylic sheets

In a more recent design for capturing conifer-feeding beetles and other forest Coleoptera, the transparent sheets which offer few visual cues are made of clear acrylic thermoplastic (acrylite) (Chenier and Philogene,

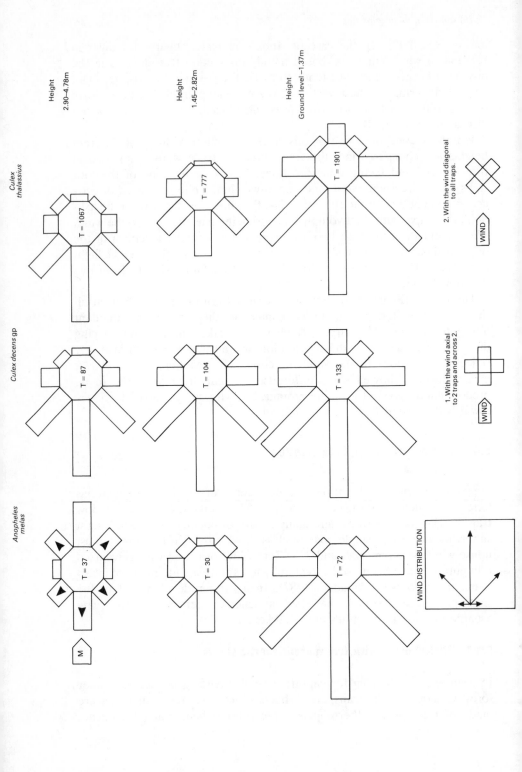

Anopheles
melas

Culex decens gp

Culex
thalassius

Height
2.90–4.78m

Height
1.45–2.82m

Height
Ground level –1.37m

M

T = 37

T = 87

T = 1067

T = 30

T = 104

T = 777

T = 72

T = 133

T = 1901

1. With the wind axial
to 2 traps and across 2.

WIND

2. With the wind diagonal
to all traps.

WIND

WIND DISTRIBUTION

1989). Two 50 × 50 cm sheets are slotted into each other at right-angles, and placed vertically, presenting a total trapping area of 1 m². An inverted pyramid made of acrylic sheets glued together forms the collecting funnel for directing falling insects into a 1 l container. A black vinyl funnel inside provides a one-way trap. A chemical lure — monoterpenes and ethanol — attached to the top of the panes attracts the beetles from a distance; beetles in flight hit the transparent panes, hung at 1.5 m, and drop.

In a series of captures involving a comparison of the window trap with two types of trap based on visual reaction to silhouettes, more than 6000 beetles belonging to 33 families were captured, and of these 15% were taken in the interceptor window trap. In many cases this appeared to be the result of random interception alone, the beetles captured not responding to aromatic baits, and this was found to apply to buprestid beetles in particular.

5.4.2 Modifications of window trap

One of the problems in studying the reactions of bark beetles and bark weevils in forest entomology is the question of whether the host tree is initially selected by a host-induced mechanism, such as by chemical or visual cues (primary attraction), or do beetles land randomly on trees, and only after landing and feeding accept the tree as a host. One difficulty in interpreting host selection mechanisms is the interference from aggregate pheromones (secondary attractants) which may promote a much stronger attraction than host odours. In order to study this question, a modified window or flight trap has been designed (Tunset *et al.*, 1988) in which the plexiglass barrier forms a 'house' (Figure 5.10) within which a test log is placed, visible from all sides. When attracted beetles fly against one of the plates, they are momentarily stunned and slide down the surface of the top and bottom plates, ending up in the collecting container underneath, which can be used with or without collecting fluid, giving the option of retaining beetles after capture. A grooved plastic tube clamped over the upper edges of the top plates effectively prevents beetles which have landed from reaching a horizontal position, the only position from which they can take off.

The log section housed in the centre of the trap is visible through the plexiglass, making it possible to include visual stimuli as part of the

---

Figure 5.9 *Direction of mosquito flight in relation to wind direction. Results with directional flight traps facing four directions at three levels. (i) With the wind axial to two traps and across two traps. (ii) With the wind diagonal to all traps. (T) Total number of mosquitoes captured at each level. < Direction of mosquito flight. □ Mean wind direction (Snow, 1977).*

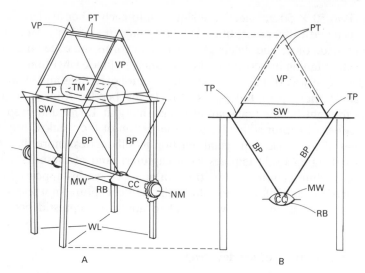

Figure 5.10 *Modified window or flight trap used for bark beetles. The described trap for test of primary attraction of bark beetles and bark weevils. A: Oblique view; B: end view. Symbols used: PT = plastic tube, VP = vertical, triangular plate, TP = top plate, BP = bottom plate, SW = steel wire, TM = test material (log), CC = collecting container, RL = removable lid, MW = metal wire, RB = rubber band, NM = nylon mesh, WL = wooden leg. (Tunset et al., 1988).*

primary attraction. By using covered and non-covered test material, visual and chemical stimuli can be differentiated.

Window traps also provide one of the several trapping methods available for studying the reactions of bark beetles to pheromones or pheromone-baited traps such as that commercially available for the bark beetle, *Ips typographus*, i.e. Ipsalure (Weslien and Bylund, 1988).

5.4.3 Combined Malaise/window trap

The window trap principle has also been incorporated in a composite interception trap designed for sampling arthropods in tree canopies in Australia (Basset, 1988). This trap consists of two sub-units (Figure 5.11). The top unit is a typical Malaise-type trap with a profile area of $0.35\,m^2$. The bottom sub-unit is a window trap with a collecting surface of $0.14\,m^2$ made of plexiglass panels, against which insects impact and fall into a collector. In five traps set at mid level in the heart of the forest crown at $25\,m$, nearly $25\,000$ specimens were collected in the first year, representing 19 different arthropod orders and 120 families. In the window trap sub-unit coleoptera made up roughly 54% of the catch, with Nematocera

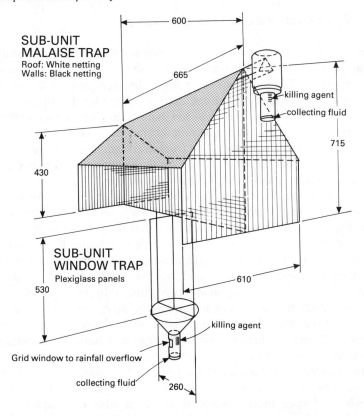

Figure 5.11 *Composite window trap/Malaise trap designed for sampling arthropods in tree canopies (Basset, 1988).*

accounting for the bulk of the insects caught in the Malaise sub-unit. In all, 89% of the insects caught were winged, the composite trap having the advantage of continuous operation day and night.

5.5 Truck or vehicle-mounted traps

The truck trap, devised originally over 30 years ago for sampling biting midges in Florida, has proved to be one of the most fruitful of the non-attractant capture techniques for many other groups of small airborne insects including blackflies (*Simulium*) and mosquitoes. It has also been used in sampling insects in general, at the 2–3 metre level after taking off, as part of boundary layer studies in Australia (Farrow and Dowse, 1984).

5.5.1 Experiments with the original design

In its original model used for trapping flying insects, the trap was simply a
screened 4-sided funnel of wood mounted on a pickup truck (Nelson and
Bellamy, 1971). The mouth of the funnel, directed forward ahead of the
windshield, was 0.6 m high and operated at a fixed level above ground of
1.8 m. The sides of the trap tapered back into fine-meshed collecting
bags. The truck was always driven at uniform rate, and at night headlights
were only used when essential. The truck trap samples very much greater
volumes of air — up to $30\,000\,m^3\,h^{-1}$ — than suction traps under similar
conditions, which usually sample less than $1000\,m^3\,h^{-1}$.

An essentially similar design of truck trap was later adopted for sampling
mosquitoes in the extensive Florida project (section 1.5.2) (Bidlingmayer,
1974) (Figure 5.12). It was not until later that the real possibilities of the
truck trap for sampling another group of small biting flies, *Simulium*,
were realized, leading to exhaustive experiments in two widely different
environments, England (Davies and Roberts, 1973) and the Sudan (El
Bashir *et al.*, 1976) and involving two quite different species. In the
Sudan, the truck trap was a much simpler and more basic design in which
a sweep net was held vertically through a slit in the roof of a saloon car
cruising at $20\,km\,h^{-1}$; there was no sophisticated or funnelling mechanism,
trapped insects simply being anaesthetized and stored.

In the English design a cone shaped net 91.5 cm wide at the mouth was
fixed on top of a van, which was usually run at the rate of 38 mph
($48\,km\,h^{-1}$). Under these conditions it was capable of sampling a very
large volume of air, up to $100\,000\,m^3$ in a half-day's operation. The insects

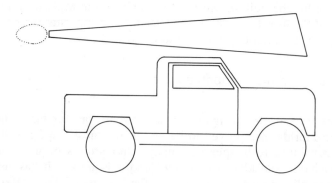

Figure 5.12 *Truck or vehicle-mounted trap for sampling mosquitoes and other
small biting flies. (Bidlingmayer, 1974).*

gathered into the net were funnelled into a pipe at the rear, which led into equipment which separated the catches by changing the collecting tubes through which the air flowed. An ingenious electric system, controllable by the driver, ensured that collections made over each kilometre could be segregated in tubes on a rotating disc.

5.5.2 Operation of truck trap

Truck traps are equally effective for both day and night-flying insects, and have been particularly useful for insects which show great crepuscular activity in flight over the period of rapidly changing illumination.

In the American studies on biting midges in Florida, which were mainly directed towards the widely distributed pest species, *Culicoides variipennis* (Nelson and Bellamy, 1971), the insects were sampled by repeatedly driving the truck over a course of about 3 km on a paved road. In a series of all-night observations one 3.2 km run was made every 15 minutes for the first six hours of the night, and a 6.4 km run made every 30 minutes in the latter six hours of the night. The results (Figure 5.13) showed peaks

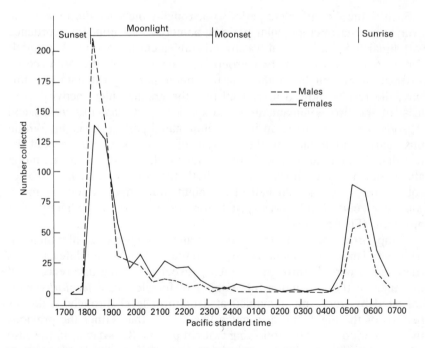

Figure 5.13 *Truck trap collections of* Culicoides variipennis *made hourly overnight at quarter moon in Florida (Nelson and Bellamy, 1971).*

of flight activity by both sexes at sunset and sunrise, the size of the peak depending on the intensity and distribution of moonlight on different nights. A more detailed breakdown of evening flight records was made by a series of consecutive 10 min runs encompassing sunset. Similar patterns of intense flight activity at dusk and dawn were recorded for several other species of Culicoides taken in these trials.

In the mosquito studies in Florida (Bidlingmayer, 1964) the practice was to drive the truck along a road parallel to, and just behind, the beach on the Atlantic Ocean for a fixed distance (2.8 km in one series and 3.0 km in another) at a speed of 15–20 mph, the net being changed at the end of each run. Regular runs of 15–20 min were made from sunset through to sunrise. The final routine adopted was according to moon phase, with trapping restricted to days marking each phase. Each night was divided into eight periods as follows:

	Evening from sunset	Night time: Dark hours	Morning until sunrise
Period	1	2 3 4 5 6 7	8

Each of the six dark night periods was roughly the same duration as the twilight and dawn crepuscular periods, namely 75–80 min. In accordance with lunar light variation, it was found convenient for periods 2, 3 and 4; and 5, 6 and 7 to form two groups characterized by the presence or absence of moonlight in the quarter moon period. Tabulated in this way, the results showed first of all that the greatest flight activity – at least of the two dominant mosquito species, *Aedes taeniorhynchus* and *A. sollicitans* – occurred during the illuminated part of the night. Within this quarter moon phase further analysis of data, taking into account the fact that the amount of moonlight will vary with the altitude of the moon above the horizon, showed that at both first quarter and last quarter collections in the moonlit half of the night were largest when the moon was near zenith, and lowest just before moonset or immediately after moonrise (Figure 5.14).

Comparisons between the two extreme conditions of illumination, namely full moon illumination throughout the night, and complete darkness (new moon period) through the night, revealed striking differences. On the night of the new moon the number of female *Aedes taeniorhynchus* collected by truck trap remained at a uniform level throughout the six periods of the night, but at a much lower level than either the previous twilight period, 1, or the following twilight period, 8, at dawn. In the case of full moon illumination through the night, a uniform level of activity – as judged by truck trap collections – was also maintained throughout the night, but at a much higher level, between six to eight times higher than

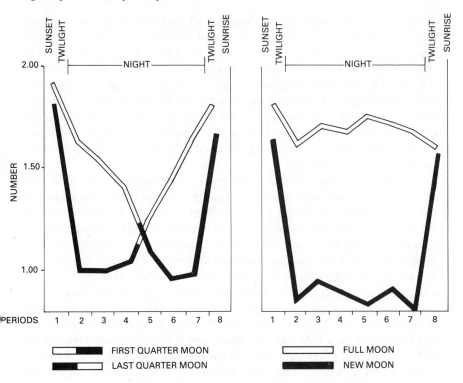

Figure 5.14 *Mean number (log) of female mosquitoes of all species taken per period by the truck trap at different moon phases (Bidlingmayer, 1964).*

the dark period of the new moon. At some points this activity approached the peak flight activity exhibited at twilight.

When these findings are applied to the earlier problem of the relative efficiency of light traps at different lunar phases (Chapter 1) it will be seen that the established fall in light trap performance at full moon periods is even greater than previously estimated due to the fact that this full moon period is actually one at which the maximum number of mosquitoes are in flight.

5.6 Tow nets

The study of insect dispersal and the factors determining the diversity of flight patterns shown by different species of winged insect has long been of absorbing interest to entomologists. An early milestone in reviewing the great amount of information already available by the late 1960s appeared just over 20 years ago (Johnson, 1969). More recent reviews (Stinner *et al.*, 1983; Scott and Achtemeier, 1987) amply demonstrate the considerable advances in knowledge since that time, but also the extent to which that progress has been linked with rapid advances in the technology of aerial trapping and sampling. These advances in turn have been greatly assisted by the development of radar and other remote sensing techniques for the detection and recording of insects in flight, particularly those involving mass movements.

Tow nets have long played a key role in such studies. The term 'tow net' embraces two very distinct trapping concepts and categories, in only one of which is the net actually towed or drawn through the air. In the other category, the so-called tow net is static, and is entirely dependent on wind movement to carry airborne insects into the net. A good example of controlled experiments involving the static form of tow net was provided by early work at Rothamsted Experimental Station in which this technique was compared with suction traps and sticky traps in its efficiency in trapping aphids (page 69) (Johnson, 1950). In this series, the net made of cotton voile was 33 in in diameter at the mouth, and 42 in long. The net was kept open by a light bamboo hoop, and a boom outside kept the net stiff. Two swivels at opposite sides of the bamboo hoop allowed the net to rotate round a vertical wire, thus keeping it facing into the wind. This model exemplifies essentially similar types of net whose use dates back as far as the 1930s, and clearly confirms that the term 'tow net' in this case is a misnomer, and that the net — normally operating 3–4 feet above ground level — should properly be considered under the heading of flight or interceptor traps, entirely dependent on natural wind flow or air currents to carry airborne insects into the net.

This dependent relation between wind velocity and trap efficiency was tested experimentally by placing the net, mounted on the cane hoop, with its opening 12 ft away from a large aerofoil fan working in a horizontal duct 18 in in diameter. The speed of the wind created by this fan was measured by velometer at nine points across the two diameters at right-angles within the cane hoop, with and without the net. This was repeated at different wind speeds, controlled by a pulley on the fan, which left the relative positions of net and fan unchanged. The mean wind speed at the net opening was expressed as a percentage of the mean wind speed through the hoop without the net, for each wind strength. This was taken to correspond with the reduction in quantity of air sampled by the net (Table 5.1).

Table 5.1

Wind speed (mph)	Air passing through net %
0.8	25.0
1.1	63.6
4.1	70.7
4.5	80.0
8.4	82.1
8.6	82.0
16.4	88.4
16.7	89.2

These figures show that at all velocities tested, the air which passes through the net is only a proportion of the amount which would pass through a similar area without a net. The net efficiency is at its lowest at low wind speeds, being least efficient under calm windless conditions, i.e. at the very time when flying insects may be most abundant. Efficiency increases with increasing wind velocity, but a point is soon reached when further increases in velocity only produce slight increases in efficiency, doubtless due to the turbulence effect produced by the net obstruction. Within the range of air speeds tested, it was never possible to attain a 90% efficiency as the absolute maximum.

In striking contrast to the use of static tow nets operating near ground level has been the increasing use of nets towed by aircraft for monitoring populations of migrating airborne insects, especially in the high altitude ranges of 5000 to 20000 ft. In such cases the term tow net is truly accurate, and involves mechanical principles which have much in common with vehicle-mounted or truck traps commonly used at ground level (page 171) As the speed of aircraft means that these tow nets are operating at much higher air velocities than would be experienced in natural winds near ground level, designs have had to be modified accordingly in order to deal with loss of efficiency when the obstruction or drag of the net produces spillage of air round the net mouth.

At the comparatively low levels of air speed encountered when light aircraft are used, roughly around $100 \, \text{km h}^{-1}$, the basic design of tow net can still be used, subject to increased strength and support. Experience with one such collapsible tow net designed to study the aerial density of the potato leafhopper, *Empoasca fabae*, in Pennsylvania (Reling and Taylor, 1984) amply demonstrates the potential of this sampling technique. The frame of the net was made of reinforced PVC tubing (215 cm in diameter) bent into a hoop to enclose an area of $0.37 \, \text{m}^2$. Nylon insect

netting formed the basis of a round-bottomed cylinder 135 cm long, reinforced with a protective collar. The net rigging, which formed an essential
part of this design, allowed the net to be deployed and retrieved while the
high-winged monoplane was in flight, and also enabled the net to collapse
on retrieval. Despite some spillage of air ahead of the net during flight,
wind tunnel experiments showed that at free air velocities between 30 and
$110 \, km \, h^{-1}$, the net only reduced the air flow by about 6%, only a minor
impairment in efficiency. In 56 flights between 500 and 2000 ft, over 1000
male and 4000 female leaf hoppers were captured, the species being
robust enough to sustain minimal damage. Some other more susceptible
taxa, including aphids, sustained damage, making identification difficult
in some cases.

At the higher velocities at which nets are towed by larger aircraft, in
the range of $200 – 400 \, km \, h^{-1}$, turbulence and spillage around the conventional net mouth make it necessary to introduce appropriate modifications to counter loss of efficiency. The basic principle is to incorporate
an expansion chamber in the collecting device which now takes the form
of a tube, narrowed at both ends, but expanded in the middle. This
allows the airflow entering the tube or trap to slow down from anything
up to $400 \, km \, h^{-1}$ to about $40 \, km \, h^{-1}$ (Gressitt *et al.*, 1961; Spillman,
1980). The very small aperture (10 cm diameter) necessary to achieve
this result at the highest air velocities of $400 \, km \, h^{-1}$ is compensated by the
greater volume of air passing through the trap at high air velocities. This
design has proved adequate for sampling low aerial densities of airborne
insects, and for ensuring that insects collected are virtually undamaged.

The problem of devising direct aerial sampling systems has been of
particular concern to Australian scientists, not only to provide quantitative
data on the densities and identities of macroinsects migrating by day or by
night at upper air levels, but also to be able to identify particular insects
whose presence and movements are being studied by radar (Farrow and
Dowse, 1984). The requirements of such a system are that (i) it should be
able to sample large volumes of air of the order of $20\,000 \, m^3$ at selected
altitudes up to 500 m, (ii) it should provide sequential sampling by day
and by night for long periods, (iii) it should be able to operate in winds
up to $100 \, km \, h^{-1}$, which commonly accompany frontal systems when
migration is frequent, and (iv) that it should be easily portable and
operable.

From previous experience of various methods used in the past, e.g. by
nets mounted on tethered kite balloons, and by nets on kites, it was clear
that these methods were not suited for the high winds involved. The
method devised was to use tow nets suspended from kites (Farrow and
Dowse, 1984). Since the aerial densities of larger insects over 1 cm in

length are relatively low, the cross-section of the net was fixed at $1 m^2$ to keep within the lifting power of the kite, and the net was made of porous material, 0.5 mm mesh, in order to reduce spillage of air from the mouth of the tow net, and retain all but the smallest insects.

In winds of $10-20 \, km \, h^{-1}$ or more the kite was launched on $100-200 \, m$ of line, which then extended to 300 up to 1000 m to make the operational height in the range of $100-500 \, m$. The net could be raised or lowered independently of the kite, and a radio operated net-closing and releasing mechanism prevented contamination of sample during the free fall of the net. Observations on insect flight in front of and inside the net indicated that few insects if any managed to avoid the net, and that none of the captured insects — including larger moths — were able to walk or fly out. The density of macroinvertebrates estimated from these net collections was found to compare favourably with those estimated by radar.

Chapter 6

Plant Pest Responses to Visual and Olfactory 'Sticky' Traps

6.1   Introduction

So-called 'sticky' traps have been widely used for trapping small winged insects, mainly agricultural pests. In very few instances however is the term strictly accurate as the treated surface traps insects in random flight or when they settle indiscriminately. In the majority of cases the sticky surface acts purely as a retentive or retaining element in a trap which depends mainly on visual, odour or chemical factors to attract the insects in the first instance. In this section, sticky traps in which the visual element predominates will be examined, while those in which odours or chemical lures are the main attractant will be described later. It should be pointed out that there are many instances where experiments with sticky traps have involved not just a single visual factor, but colour and odour as well. In addition, a silhouette factor may be involved comparable to that which has long been known to operate in many of the visual traps used for studying biting insect pests of man and domestic animals.

6.2 Response to colour and coloured sticky surfaces

6.2.1 Experiments under greenhouse conditions

Traps based on response to colour have been widely used in crop pest management. Most of the basic problems encountered in studying insect response to coloured sticky surfaces, both under greenhouse conditions and on fruit crops in the field, are well illustrated in the following example.

(a) Greenhouse whitefly
Greenhouse whitefly *Trialeurodes vaporariorum* are well known pests of tomato crops, controllable under certain conditions by introducing parasitic Hymenoptera. For such control to be effective, it is essential to devise monitoring methods which can detect extremely low populations of whitefly, before they flare up beyond parasite control (Gillespie and Quiring, 1987). Also important is the fact that adults and immature greenhouse whitefly are highly clumped in their distribution, requiring special techniques. These whitefly are well known to be attracted to various hues of yellow (Vaishampayan *et al.*, 1975) and accordingly bright yellow plastic strips, 60 × 4 cm, coated with sticky adhesive were used in these experiments. These tests had two objectives; first, to check the efficacy of these coloured traps by comparing the catch with population estimates based on weekly counts of adults on plants, made on the same day the traps were coated, and second, to study the effect of trap density.

On the question of trap efficiency, over the 13 weeks of observations — in which there were approximately three generations of whitefly — both capture methods showed the same general trend (Figure 6.1). However, the peak recorded on traps in weeks 4–6, i.e. 2 to 2.5 per plant was well in excess of the similar peak in plant catches, which did not exceed 0.4. This was attributed to the fact that at those population levels, the traps exerted a controlling effect, suppressing the population on the plants themselves.

To assess the influence of trap density three experimental blocks were used in which the trap density per 180 plants was 2 (A), 5 (B) and 10 (C). Each week counts on plants and on traps were converted into numbers of

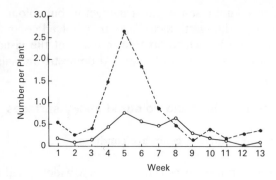

Figure 6.1 *Numbers of whitefly adults on plants (solid line) and weekly catches of whitefly adults (broken line) in a greenhouse tomato crop (Gillespie and Quiring, 1987).*

adults per plant. The results (Figure 6.2) showed that as trap density increased, there was an increase in the proportion of adults trapped, relative to those on plants. At a density of ten traps per 180 plants, the population of adults did not exceed 0.1 per plant, whereas at lower trap densities, whitefly populations were considerably higher. Only at this trap density was there a significant relationship between trap catch and population on plants. Whitefly adults would not be expected to be found in traps at densities between 0.01 and 0.1 adults per plant, although in practice the first pocket of infection in a greenhouse would be detected at a lower level.

A trap density of at least one trap per 18–20 plants, i.e. ten per 180–200 would be required to monitor adult whitefly populations on a tomato crop. At lower trap densities, the whitefly population fluctuates more markedly and is consistently higher than at high trap densities. This is attributed to the trapping-out or suppressant effect on the whitefly population produced by the higher trap densities.

(b) Responses of greenhouse whitefly to different colours
In another investigation, the response of greenhouse whitefly to coloured traps was studied in the context of its role as a world-wide pest of greenhouse-grown ornamental and vegetable crops. The development of resistance by this insect to many insecticides emphasized the need for an integrated control programme based on trapping adults (Webb *et al.*, 1985). Starting from the well established fact of the attraction of greenhouse whitefly to yellow–green portions of the spectrum (Vaishampayan *et al.*, 1975), a comparison was made between six colours in addition to

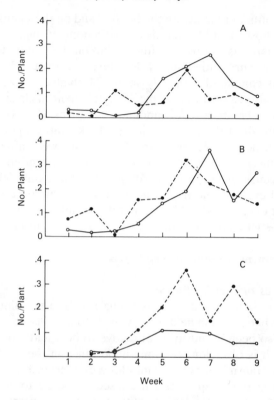

Figure 6.2 *Numbers of whitefly adults on plants (solid line) and weekly catches of whitefly adults (broken line) in a greenhouse tomato crop. (a) Trap density of two per 180 plants; (b) five per 180 plants, and (c) 10 per 180 plants (Gillespie and Quiring, 1987).*

black and white. These coloured cards, treated with a standard adhesive Tick-Tack were placed randomly around a 60 cm radius circle, 30 cm above the bench in a greenhouse. The results confirmed the superior attraction of yellow over other colours, and showed that while green and orange were less attractive, they in turn were more attractive than white, violet, blue, red or black. The attraction to bright yellow traps was even more marked than attraction to the natural green leaves themselves, and this fact formed the basis of a successful trap-out strategy of control.

The number of traps necessary to achieve this trapping-out effect was found to depend on the host plant, and this was taken into account when the 12 sticky yellow boards were placed among the plants. Greenhouse whitefly adults were then released into the greenhouse, and eight days afterwards counts were made on the boards, and on the plants. In tomato, the favoured host plant, 12 boards placed among 16 plants yielded 3621 flies on the boards as compared with 882 still on the plants. In the case of the greenhouse chrysanthemum, a less favoured host, 12 traps yielded 2575 adults as compared with only six on 24 plants.

Tests were also carried out on the sticky element of these traps. These showed that in a comparison of candidate adhesives, none of these caught all the whitefly all of the time. Undiluted Tack-Trap caught over 90% of the whitefly attracted to yellow boards for at least a week after application, but other adhesives, effective on the day of application, rapidly lost efficiency because of evaporation. With still others, only 50% efficiency was recorded even on the day of application. Greenhouse whitefly have powdery wax over wings and body, and this enables them to escape from a number of sticky materials that entrap most other insects.

6.2.2 Experiments in outdoor conditions

(a) Responses of beet leafhopper
Essentially the same methods of assessing response to coloured traps or surfaces have been used in field trials with allied pest insects. For example, yellow has also been found to be attractive to beet leafhopper, *Circulifer tenellus*, which is of major economic importance in the USA because of its ability to transmit several plant diseases (Meyerdirk and Oldfield, 1985). Trials were set up to test the insect response to coloured sticky traps in order to develop an effective tool for monitoring population densities, and to determine flight activity in citrus and vegetable crops, as well as in wild host areas. Another objective was to determine the trap height that would maximize capture of adults.

Coloured translucent plastic cards 12.7 × 17.8 cm, coated with tangle-foot, were exposed vertically on stationary rods inserted in a wooden frame (Figure 6.3). On each frame six cards with different colours were tested randomly, yellow, green, blue, red, white and clear (transparent, no colour). Cards were exposed for 2 week periods for each test. A further series of tests with different hues of yellow, using coloured poly-ethylene discs 15.5 cm in diameter, was carried out for the comparison of yellow, fluorescent yellow, gold and clear. The colour preference experiments showed that yellow with a dominant wavelength of 570 nm was significantly more attractive to male and female leafhopper than the other colours, and that within the different hues of yellow there was no preference.

In tests to determine optimum height of traps, translucent plastic cards 13.8 × 22.7 cm were set up at heights from ground level of 0.3, 0.6, 0.9,

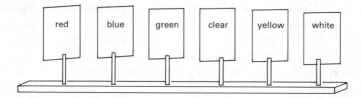

Figure 6.3 *Arrangement of colour cards in trap response studies on beet leafhopper (Meyerdirk and Oldfield, 1985).*

1.5, 2.1 and 2.7 m. These showed that the ground level traps caught more males and females than any other traps; very few adults were caught at or above 0.6 m.

(b) Responses of citrus blackfly
In the case of an allied homopteran insect pest in the USA, the citrus blackfly, *Aleurocanthus woglumi*, yellow sticky traps provide one of the main survey methods (Hart *et al.*, 1978). Again it was shown that yellow with a reflectance of 550−555 nm had the greatest attraction. For this test species the possible influence of trap shape was also tested using square, circular, rectangular and equilateral triangle shapes, spray-painted with fluorescent yellow, and treated with Tack-Trap. Different shapes all proved to be equally attractive (Meyerdirk *et al.*, 1979)

(c) Responses of citrus psyllid
Response to yellow traps has received particular attention in the citrus growing areas of South Africa where the psyllid *Trioza erytreae* is a vector of citrus greening disease (Samways, 1987a). Practical considerations emphasize the need for a low cost trapping method capable of predicting population upsurges. The citrus psylla is positively phototactic to yellow−green wavelengths (Urban, 1976) and is an active flying insect over several hundred metres, enabling it rapidly to invade areas of flushing vegetation.

A comparison was made in one of the orchards between trap catches and actual counts of adults on mature citrus trees, after the manner of the greenhouse whitefly tests described above. Traps comprised a sheet of perspex 28 × 25 cm, on one side of which was stuck a 25 cm square of fluorescent yellow self-adhesive tape covered with a PVC transparency, the centre of which was smeared with clear polybutene. The square was marked with small squares for counting *Psylla*. Traps were hung at 1.5 m (Samways, 1987a,b). The trap unit or trap set comprised two groups of three traps. Under favourable conditions the population per trap-set per week occasionally exceeded 100, but was generally much lower. At high population levels trap catches were no more sensitive than visual counts, but at low levels the traps proved to be more sensitive. Visual counts had the added disadvantage of taking much longer to check. The results suggested that a threshold of more than two individuals per set of three traps in two consecutive weeks is a sensitive indicator of potential upsurge in *Psylla* population. As adult *Psylla* must be about 21 days old to be infective, there is ample time to respond to trap results.

The particular type of yellow, with a high reflectance in the yellow combined with a low reflectance in the blue, was provided by 'Saturn yellow' which caught the highest number of *Psylla*, and also proved attractive to several other citrus pests.

6.2.3 Effect of angle of landing surface

One sampling problem familiar to many entomologists faced with the task of monitoring pest Diptera on vegetable crops is to decide the relative merits of sticky traps versus water traps. Both techniques have known disadvantages. For example, vertical sticky traps are more liable to catch non-target flies such as anthomyids and blowflies, while water traps are prone to trap beneficial insects such as syrphids.

In order to obtain more information about the responses of these pest flies to sticky traps, experiments were designed to find out if they have a preference for landing on a particular plane. The three main pest Diptera in question were the carrot fly, *Psila rosae*, the turnip root fly, *Delia floralis*, and the onion fly, *Delia antiqua*. Yellow sticky traps were aligned at the eight different angles of an octagon (Figure 6.4). Each trap consisted of a 10 × 20 cm sheet of acrylic plastic, yellow in the case of the carrot fly and fluorescent yellow for the other species. The traps for the carrot fly were tested in a 1000 m² plot, while in the case of the other two species — which had to be reared — flies were released into cages 6 × 3 × 2 m erected over crops of swedes and onions respectively (Finch and Collier, 1989).

The species of *Delia* preferred to land on the horizontal surface (trap 1), or those with the sticky surface facing upwards, and inclined at 45° to the vertical (traps 2 and 8). In contrast, most *Psila rosae* were caught on the two traps inclined at 45° to the vertical, but with the sticky surface facing downwards (traps 4 and 6). This preference for the underside of angled sticky traps has the advantage, in monitoring, that this sticky surface is more protected in all weathers.

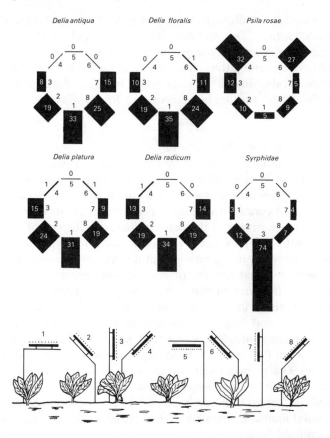

Figure 6.4 *Percentage of carrot fly,* Psila rosae, *and four species of* Delia *caught on yellow sticky boards aligned in one plane, but orientated in eight different directions (Finch and Collier, 1989).*

6.3 Responses to two- and three-dimensional sticky traps

6.3.1 Introduction

Under the wide general term of sticky traps in which the prime attraction is a visual one, the traps described so far have been two-dimensional, in the form of coloured cards or strips. With some insect traps a further element has been introduced in the form of three-dimensional coloured sphere, coated with adhesive. Onto this in turn may be added a chemical or odour attractant. This makes it difficult to break down such trapping techniques into strict categories, because two or more factors may be involved in determining trap effectiveness. This is perhaps best illustrated by the considerable amount of work which has been done on a major fruit pest in the USA, the apple maggot fly, *Rhagoletis pomonella* (Jones, 1988; Stanley *et al.*, 1987).

(a)   Development of 'artificial apple'

Earlier observations in the laboratory established the value of an 'artificial apple' in the analysis of fly response to the natural host fruit (Prokopy *et al.*, 1972). This artifact took the form of a 25 mm diameter hollow orange-coloured ceresin wax dome. These experiments showed that few flies of either sex arrived at the ceresin domes — the site of assembly for mating — until the flies were 7–8 days old. From then on there is a progressive increase in assembly, mating and oviposition. Field observations confirmed that the site of male/female assembly for mating was exclusively on the fruit of the larval host plant (Prokopy *et al.*, 1971). Further observations in the field showed that activity is mainly determined by temperature and light. As the day progresses, both sexes begin to leave their resting sites on leaves and move to the fruit. During the afternoon, when flies are present in large numbers, more than 80% of all females and 90% of males were on the fruit. Towards evening flies begin to leave the fruit, and disappear from view. Laboratory observations confirmed the importance of visual cues in the location of both the fruit and potential mates, in showing that in darkness there is no assembly of flies on artificial fruit.

(b)   Attraction of red sticky spheres

The attraction of red sticky spheres was investigated by comparing them with yellow sticky panels (Kring, 1970). Yellow panels or yellow boards attract flies as a possible food source, while the red spheres act mainly as a mating and oviposition site. The attraction of these two was examined both separately and in combination, with the main object of finding the most effective lure for dispensing a chemosterilant. Among the attractants offered were yellow panels with a round red 7.5 cm spot in the centre of each side, and yellow panels with a round sphere nailed to the centre, all objects being coated with Stickem. A record was made of the exact location of flies trapped on panels with and without spheres. These trials showed that a red spot painted on a yellow panel was unattractive, in contrast to the high attraction of a red hemisphere in the centre of the panel.

(c)   Combination of sticky spheres and chemical lure

In another experiment, attraction to the visual stimulus of sticky spheres was supplemented by means of chemical lures (Reissig, 1974). Artificial apples, coated with Stickem, had two holes punched at opposite ends to allow incorporation of baits within the apple. These experiments showed that baited sticky apples were more attractive than non-baited ones. From this it appears that the apple maggot fly is directed by both visual and olfactory stimuli.

Considerable attention has been given to the positioning of traps within trees (Reissig, 1975). In field trials, three canopy position variables were examined, canopy radii, height above ground and compass direction. In addition to a comparison between a sticky yellow card and a red plastic sphere coated with Tanglefoot, a third trap based on chemical attractant (a mixture of ammonium acetate and yeast) was included in the trial, in an orchard where tree heights ranged from 3.0–3.5 m.

All of these traps caught more flies higher in the trees (Figure 6.5). Traps 1.2 m above ground were ineffective; for example, the chemical-baited trap caught an average of 96.8 flies per trap during the season at 3 m, compared with 40.8 flies per trap at 2.1 m and 13.2 flies at 1.2 m.

Three canopy positions were selected for further study: inside the canopy on a main branch 0.3–0.6 m from the tree trunk; middle of canopy radius; and outside canopy 0.3 m or less from the outside foliage. The trends shown by all three traps demonstrated that they were least effective in the outside canopy, and most effective in the middle canopy, indicating that adequate shade is a more important factor than surrounding foliage in ensuring maximum efficiency of traps. The results also emphasize the importance of precisely standardized canopy position in establishing reliable data on population density and population fluctuations, as well as in comparing of different trapping systems.

Catches on coated red spheres have been measured in relation to positioning of adjoining natural fruit in the tree, e.g. fruit above trap only; fruit solely below trap; and fruit both above and below (Drummond *et al.*, 1984), but none of these were found to have a significant effect. However, the distance of apples from the trap is evidently important. Fruit too close to a red sphere, e.g. less than 0.25 m, masks trap effectiveness, decreasing trap capture, but the absence of fruit within a 1 m radius of the trap also reduced effectiveness. Fruit at intermediate distances of 0.25–0.5 m provided the best conditions for trapping by sticky red spheres.

(d)   Combinations of red spheres and apple volatiles

The use of red spheres for trapping apple maggot fly was enhanced by the identification of the volatile chemical from 'Red Delicious' apples (Fein

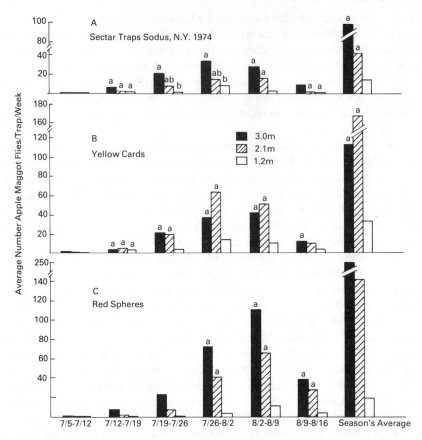

Figure 6.5 *Relative effectiveness of three types of apple maggot traps hung 1.2, 2.1 and 3.0 m above ground in apple trees (Reissig, 1975).*

et al., 1982). Red spheres baited with this volatile were found to capture more males and females than unbaited spheres which provided visual stimuli alone (Reissig *et al.*, 1982). This led to a more critical comparison between baited and unbaited spheres, as well as baited yellow triangles (Reissig *et al.*, 1985), carried out through an entire season in different orchards in New York. Three types of trap were used: (i) a dark red wooden sphere 8.5 cm in diameter coated with bird Tanglefoot; (ii) a sphere identical to the above, but baited with 100 ml of a mixture of synthetic apple volatile released from a polythene cap, and (iii) Pherocon AM pre-baited yellow triangle. Traps were suspended along the outside edge of the tree canopy, 2 m above ground.

The results in 1982 showed that the baited spheres caught considerably more flies than the other two types in five sampling intervals. In the first two intervals, when catches were lowest, the baited spheres caught 89% of the flies. Later in the season, when flies became more abundant, baited spheres caught from 2.5 to 4.4 times as many flies as unbaited ones, and were also 52 times more effective than the panels. Under field conditions the sustained-release polyethylene bottle dispensed the volatile blend for at least six weeks. In practical terms the trap proved sensitive in detecting the low densities of apple maggot adults causing only traces of fruit damage (Reissig *et al.*, 1985) and flies were detected in every case in which apple damage was observed in the 83 trees monitored.

Eighty percent of apple maggot oviposition occurs during the afternoon when adults tend to be evenly distributed throughout the canopy. The volatile baited spheres do not cause major shifts in this pattern of oviposition within the canopy. Flies tend to concentrate in the top of the trees in the early morning and in the evening (Prokopy *et al.*, 1972) and this is reflected in increased trap catch in the upper portion of the canopy.

6.3.2 Response of mediterranean fruit fly, *Ceratitis capitata*, to model fruits

Work in Hawaii and elsewhere has shown that, like the apple maggot fly, the mediterranean fruit fly also responds visually to model fruits or coloured spheres (Nakagawa *et al.*, 1971, 1978), and this question has been further investigated in its original Old World area of distribution. In experiments carried out in Greece, plastic spheres 7.0 cm diameter (the preferred size) were painted with seven different enamels, black, red, orange, white, green, blue and yellow, and coated with Tanglefoot. In this area the natural hosts plants are citrus, olive and figs. In a 45 day experiment 9758 flies were captured on all spheres; yellow proved to be the most attractive colour, followed by orange, black, red and green, with white and blue the least attractive. Flies of both sexes responded positively to certain wavelengths, especially the hues reflecting maximally between 570 and 580 nm. The brightness of the spheres was evidently not responsible for the observed colour preference (Katsoyannos, 1987; Katsoyannos *et al.*, 1986).

Response to coloured spheres is primarily a food-seeking response to ripe fruit-type stimuli, rather than to ovipositing-site stimuli, despite the fact that 71% of the captured flies are females. In the laboratory ready-to-oviposit females show quite a different colour preference from that observed in the field, blue, black and red being most preferred for oviposition, and yellow having the lowest preference. These results, and their interpretation, differ from those obtained in Hawaii on coffee trees as host plants, and in Brazil with peach trees as host plants (Cytrynowicz *et al.*, 1982; Nakagawa *et al.*, 1978). In the former area *Ceratitis* was most attracted to black and yellow, while in Brazil the preferred colours were red and black, followed by yellow. It seems likely that these apparently conflicting reports on colour responses of the same species, *Ceratitis capitata*, must be related to the evolution of different preferred host plants in these widely separated areas, and to the dominant colours associated with the respective hosts.

6.4 Traps based primarily on synthetic chemical attractants for males

6.4.1 Background to chemical attractants. Studies on fruit flies (Tephritidae) in the USA and Hawaii

The basic attraction of adult apple maggot fly to synthetic volatiles, introduced in this chapter under the heading of coloured visual traps, could have just as appropriately been discussed under a much wider heading of reaction of fruit pests to chemical attractants. With that classification, the emphasis would be on the chemical itself rather than visual response to the trap.

Chemical attraction in fruit pests is well illustrated by the great amount of work carried out over many years on the oriental fruit fly, *Dacus dorsalis*, and the mediterranean fruit fly, *Ceratitis capitata*, in the USA and Hawaii. Chemical attractants such as methyl eugenol were used for many years to attract oriental fruit fly in Hawaii, Mexico and elsewhere (Steiner, 1957). The discovery of this pest insect in Florida in 1956 led to intensive studies on attractants, among which angelica seed oil proved to be markedly more effective. This in turn led to the design of improved traps, with the chemical impregnated on cotton rather than in emulsifiable form (Steiner, 1957). The trap designed (Figure 6.6) took the form of a horizontal polystyrene cylinder $5\frac{1}{2}$ in long by $3\frac{1}{2}$–$4\frac{1}{2}$ in wide. The lure (angelica seed oil) was initially impregnated on cotton, with DDVP insecticide as the lethal element. More than 60 000 of these 'Steiner' traps were used in the eradication programme in Florida.

6.4.2 Development of synthetic attractants for Mediterranean fruit fly

In this successful campaign, an important part was played by Siglure, a new synthetic attractant for Mediterranean fruit fly (*sec* butyl-6-methyl-3-cyclohexane-*i*-carboxylate) (Beroza *et al.*, 1961). In practice, certain

Steiner
Jackson
McPhail

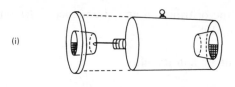

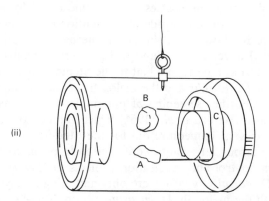

Figure 6.6 *(i) Steiner trap used in experiments on fruit flies (Steiner, 1957). (ii) Showing location of wicks of different lengths. A. 1½-inch wick used for angelica oil. B. 3-inch wick in form of loop. C. 6-inch wick inside trap cover.*

commercial batches of Siglure gave inferior performance, and this led to investigations on other isomers, which were laboratory and field tested for attractance and lasting quality. From this emerged first, Medlure, followed by an even more effective attractant, Trimedlure. This new lure, combined with a toxicant, is a valuable control measure directed towards the male flies. When used in Steiner traps, the male fruit flies do not feed as readily on these synthetic lures as on angelica seed oil, but they congregate closely around treated wicks or surfaces. If the concentration is too high, the flies are repelled.

In comparison with two other types of trap in Hawaii, the Steiner trap proved to be the most sensitive (Nakagawa *et al.*, 1971). For sustained maximum effect it was sufficient to treat the wick in the plastic trap every 3−4 weeks, but if no further retreatment was carried out, there was a noticeable decline in catch after 16−18 weeks. Steiner traps baited with Trimedlure continue to be used in Hawaii in order to provide the essential

information about the distribution and abundance of Mediterranean fruit
fly (Harris and Lee, 1987).

6.4.3 Oriental fruit fly and its allies

(a) Attractants for Oriental fruitfly, Dacus dorsalis
Dacus dorsalis is only one of a large group of tephritid or fruit flies, the
Dacini, which infest a wide range of fruits in many tropical and subtropical
countries. This particular species — introduced briefly earlier in this
section (6.4.1) — attracted considerable attention following the disclosure
of its presence in Hawaii in 1946. As in the case of the Mediterranean
fruit fly, methyl eugenol was found to be a very attractive natural lure,
and this was used as the original bait in the Steiner traps which proved as
successful with *D. dorsalis* and the melon fly, *D. cucurbitae*, as they had
with 'Med-fly'. Further screening of potential synthetic male attractants
yielded Cue-lure (Beroza *et al.*, 1961) which fulfilled the same function
for Dacus species as Trimedlure did for the Med-fly.

When these strong attractants were used in traps, in combination with
insecticide, in control and eradication projects in Hawaii, reductions in
numbers of *D. dorsalis* by 90% were obtained using methyl eugenol, and
over 90% with the melon fly *D. cucurbitae*, using Cue-lure (Cunningham
and Steiner, 1972).

(b) Australian work on specificity of male lures
These chemical attractants are of considerable interest to workers in
Australia and the South Pacific area where over 80 pest species of Dacini
fruit flies occur (Drew and Hooper, 1981). In their experiments Steiner
traps were used with a three-fold purpose: (i) in population detection to
assist with the timing of spraying control programmes in orchards; (ii) the
assessment of populations in ecological studies; and (iii) quarantine surveys
aimed at detection of introduced exotic species.

Of the 55 species of *Dacus* tested, 39 responded to Cue-lure, and 16 to
methyl eugenol; none responded to Trimedlure (Drew, 1974; Drew and
Hooper, 1981). No species were found to be attracted to both Cue-lure
and methyl eugenol.

As all these chemicals are essentially male attractants, they are considered
to act as parapheromones for fruit flies. More attention has been directed
to the question of their exact effect on the female population when used
in trapping systems. As in the case of the Med-fly, both sexes of various
species of *Dacus* are attracted to coloured sticky traps, but if male
attractant is added to these, there is a substantial fall in the number of
females captured (Hill and Hooper, 1984). Experiments were designed

to examine this quantitatively using four tephritid species and the male lures specific to them (Hill, 1986). Of these four, *Dacus tryoni* and *D. neohumeralis* males are normally attracted to Cue-lure; *D. cacuminatus* responds strongly to methyl eugenol, and *Ceratitis capitata* (Med-fly) are strongly attracted to Trimedlure.

The experiments were of two kinds: a comparison of flies captured on yellow–green coloured sticky traps with or without male lure, and a comparison of flies captured on plastic traps baited with protein solutions (active yeast) with or without male lures. The results showed that over each experimental period, the presence of male lures reduced female fruit fly captures consistently, whether they were combined with attractant colours or a food bait. The addition of synthetic lures served to nullify the attraction to colours by the females. The repellent action of male lures is strong enough almost entirely to counteract the attractive properties of the protein bait for *Dacus* females, and to decrease considerably its attraction for *Ceratitis capitata* females.

(c) Recent developments in trapping the olive fruit fly, Dacus oleae

At the present time control of *Dacus oleae* in the Mediterranean region is carried out by a combination of insecticides and attractants. Like many other fruit flies of this group, this species was found to be attracted to yellow colour (Economopoulos, 1977, 1979) and this led to the possibility of using coloured traps for monitoring purposes. Unlike most fruit flies, however, *Dacus oleae* had a long history of chemical attractants based on protein hydrolysates and ammonium carbonates. There was also a long tradition of using these baits in a simple clear-glass trap, the McPhail trap, originally designed over 50 years ago for attracting Mexican fruit fly by means of fermenting sugar solution (McPhail, 1937) (Figure 6.7). In Greece at least the McPhail trap remains the standard against which all other candidate attractants are tested.

When baited with proteinaceous or ammonium type solutions, McPhail traps attracted greater numbers of *Dacus oleae* than the yellow-coloured trap (Economopoulos, 1979), and other workers have confirmed this by showing that yellow traps attract only one-tenth the number taken in baited McPhail traps. From this result arose the idea of combining the best features of these two attractant elements in a new trap, more robust than the glass McPhail type (Zervas, 1982). This new trap took the form of a translucent PVC bottle, 30 cm high and 9 cm in diameter, and with an outer surface of 825 cm^2. The attractant solution which filled the bottle was evaporated through a roll of filter paper. When ammonium sulphate was used as bait, the new trap took higher catches than the McPhail trap, but was less effective in the case of other attractants — Buminol, a

Figure 6.7 *McPhail's invaginated glass trap.*

protein hydrolysate, 'Dacus bait', and Entomosyl. When the outer surface of the new trap was painted yellow, the catch was markedly increased over the McPhail trap in three out of the four attractants, ammonium sulphate, Buminol, and 'Dacus bait' by 5.55, 2.13 and 3.16-fold respectively.

Most trapping systems for olive flies have proved to have certain limitations. Yellow traps, for example, have a strong adverse effect on the beneficial fauna of the olive ecosystem, and sticky traps are liable to lose their efficiency in retaining insects which have been attracted to land on the surface. Attention is now more concentrated on combining in the same trap a sex attractant — the most effective male attractant — and ammonia-releasing dispensers, which are the most effective female attractants, and which are attractive to males as well. The object of this type of investigation was to develop a method of control by mass trapping rather than simply a device for monitoring (Haniotakis *et al.*, 1986).

The traps took the form of a plywood rectangle — dipped in insecticide — and baited with two dispensers, one containing a mixture of sex pheromone, and the other containing ammonium bicarbonate, a food attractant for both sexes. As this trapping system was aimed purely for practical control purposes, there is as yet no precise information about how these flies respond to this combination of attractants.

Chapter 7

Responses of blood-sucking flies to visual traps

7.1 Introduction to 'animal-model' traps

Visual attraction to the human or animal host has long been recognized as a major factor in the reactions of day-time biting flies seeking a blood meal. This is particularly evident in the case of the larger two-winged flies such as horse flies and tsetse flies, whose flight can be observed and

followed, but it also applies to other flies which are active by day such as
blackflies, *Simulium*, as well as those species of mosquito which attack
their hosts by day rather than by night.

In the field of tsetse fly behaviour in particular, visual attraction has
long been recognized as a major factor determining the success or otherwise
of the long succession of unbaited trap designs devised over the years,
many of which, e.g. the 'Harris' trap and the 'Chorley' trap, are named
after their originators. In the last 20 years the continuing problem of
tsetse response to both animal baits and to unbaited traps has been the
subject of intensive investigation by new critical experimental methods,
particularly in Zimbabwe, which has introduced new methods of analysis
of trap response of these and other blood-sucking insects. Before dealing
with these more recent developments it would be instructive to examine
one particular example from the earlier tsetse research era, in which a
basic design of unbaited trap devised for tsetse flies in Africa was later
extended to horse flies (*Tabanus*) as well, not only in Africa but also in
the USA. The basic problems encountered by those original workers
were precisely the same as those which have been re-examined so successfully by a later generation of research workers in Zimbabwe and other
parts of Africa.

7.1.1 Trials with the Morris trap in West and East Africa

This design, originally named the 'animal trap' by its devisers (Morris and
Morris, 1949) and often referred to as the 'Morris' trap, will be referred
to in this review as an 'animal-model' trap in order to clarify the fact that
no live bait is involved. The trap was originally designed to resemble a
small host animal about the size of a goat for capture of the West African
species of tsetse, *Glossina palpalis* and *G. tachinoides*, but its use was
later extended to the game tsetse of East Africa, in particular *G. pallidipes*
(Morris, 1960). The most important feature of the trap was the cylindrical
shape of the body constructed of hessian cloth, which was open along the
ventral surface, producing highlights and shadows from the curved shape.
Flies attracted into this dark opening along the undersurface were directed
from inside to a slit along the upper surface of the body, over which a
cage was superimposed. A non-return device ensured the capture of all
tsetse entering the cage.

When these traps were sited carefully, with due regard to visibility from
a wide angle, and presence of natural hosts in the vicinity, they proved to
be very productive. Comparative trials in West Africa and in East Africa —
where a slightly larger 'animal' was found necessary to attract the species
there — showed that for most of the year animal-model traps were much

more productive than fly-rounds and fly-boy catches being carried out at the same time. Even more important was the fact that female tsetse made up a higher proportion (up to 82%) of the Morris trap catch. The two sexes of tsetse are known to emerge from puparia in approximately equal numbers but, as females have a considerably longer life span than males, a wild population of tsetse comprises 70–80% of females. With all the species of tsetse tested, the proportion of females in this design of trap was very close to that estimated in the natural state, and this would appear to indicate that this trap sample is a good representation of the whole population. This is in striking contrast to the long-established method of capturing tsetse attracted to live human bait by means of the fly round, which in the case of the savanna species of tsetse consistently yields catches in which males predominate, up to 90%.

As the animal-model trap catches a much more representative sample of the tsetse population, it might be expected to be a more accurate pointer to determining normal behaviour patterns. However, nothing in tsetse research is ever as simple as first appears; when the Morris animal-model trap was compared with live cattle bait in trials in East Africa, the traps were found to be sampling a different proportion of the population from that attracted to cattle. The oldest females and those most advanced in pregnancy were more readily taken in traps. The trap, despite its animal-like appearance and design was not simply functioning as a substitute animal, but must also have been providing additional attraction to these flies, especially females, seeking a shaded resting place in the hot bright period of the day. A further difficulty is that in the particular study area in West Africa where the original model was based on an animal about the same size as a goat, the live animal itself, namely the goat, was not attractive to *Glossina palpalis*.

7.1.2 The animal-model (Morris) trap for tabanids

Later, attention was extended to the potential of this capture technique for African tabanids (Morris, 1963). Until that time surveys of these flies had been carried out as an extension of routine tsetse fly surveys based on hand-catching with nets by teams of fly collectors. For this purpose a larger animal-model trap was used, 4 ft long and 4 ft high as compared with the normal 2 × 2 ft for tsetse. Over 50 such traps were used in a survey, mainly along river banks in forested areas. Of 36 species of tabanid recorded, many were taken both in the traps and by hand-catching of flies attracted to human bait. A few species were found in traps alone, while others, such as *Haematopopa*, taken by human bait, were not recorded in the animal-model trap.

An animal-model trap, based on the African design, was tried out in the course of intensive investigations on comparative trapping methods for American tabanids (Thompson, 1969; Thompson and Pechuman, 1970). Under those conditions the model proved to be less effective than the more widely used Manitoba trap (Section 7.4) which captured all known species in the area. However, the Morris trap proved effective for one important tabanid species, *Tabanus quinquevittatus*, a widely distributed livestock pest. In contrast to the experience with tsetse flies in which both sexes were taken in this type of trap, only female tabanids were caught, the non-bloodsucking males evidently not being attracted.

7.2 Silhouette traps for blackflies (*Simulium*)

7.2.1 Experiments with three-dimensional trap design in Canada

Essentially the same principle involved in the animal-model traps for tsetse and tabanids has been developed, quite independently, in blackfly (*Simulium*) studies in Canada (Fredeen, 1961), in connection with *Simulium arcticum*, a widespread pest of cattle and livestock. Although these designs were referred to as 'silhouette' traps, most of them were in fact three-dimensional, and come in the category of animal-models. Three types were originally tested: the largest, cow trap, 4 ft high and 5 ft long and 2 ft wide, was set up on four legs; the 'sheep' silhouette was 18 in high, 2 ft long and 1 ft wide. In addition, a pyramidal design was tested. The success of the cow design appeared to indicate that efficiency is a direct function of surface area and size of opening, rather than shape. This trap sampled the attacking female blackfly population alone, and unlike the responses of tsetse to a similar type of animal-model, was not attractive to females seeking shelter, or to those in intermediate stages of blood digestion.

The possibilities of this basic design of 'cow silhouette' have more recently been explored with another Canadian pest species of blackfly, *Simulium luggeri* (Mason, 1986). Because of this species' habit of swarming around the head of cattle, the model was provided with a head, which also had a separate opening on the underside, and a separate collecting chamber (Figure 7.1). Twenty traps were distributed 100 m apart, one trap of each pair being provided with chemical attractant in the form of CO_2 from dry ice or from a cylinder. The results from both baited and unbaited models showed that the numbers caught on the body greatly exceeded those on the head. Furthermore the CO_2 based trap proved to be more than ten times as attractive as unbaited traps, with a total catch

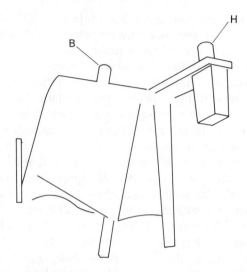

Figure 7.1 *Construction of animal-model or cow silhouette trap for blackflies. H. 'Head' collection chamber; B. 'Body' collection chamber (Mason, 1986).*

of 4826 out of 5272. However, the results could not simply be interpreted as an overwhelming visual preference for the body of the trap over the head, and appeared to be much more closely associated with the wide difference in the size of the ingress openings to these two components, that of the body being $0.927\,m^2$ as compared with $0.07\,m^2$ into the head. This endorses the experience from the original design (Fredeen, 1961) that surface area and size of opening mainly determine the efficiency of this trap in catching *Simulium*.

The additional stimulus provided by the chemical attractant, CO_2, is in keeping with experiences with many other type of trap, both interceptors such as the Malaise trap and visual traps like the Manitoba, which were originally designed without this additive, but in which the inclusion of CO_2 is now standard practice.

7.2.2 Observations on two-dimensional (true silhouette) traps in Canada and Europe

The use of true two-dimensional silhouettes has also been tested with *Simulium*. In some earlier studies in Europe (Wenk and Schlorer, 1963) a range of animal and bird silhouettes were used, ingeniously constructed in such a way that the actual silhouette of each animal could be varied as to position of head, ears, tail and other projections. Those observations showed that *Simulium* were not only attracted to different model hosts, but also to particular areas of the simulated animal.

The potentialities of this line of study were further developed in Canadian studies on *Simulium euryadminiculum*, whose biting habits are entirely specific to a particular bird species, the common loon, *Gavia immer* (Bennet *et al.*, 1972). In experiments with visual targets, both true silhouettes and three-dimensional designs were used. The attraction of these visual traps was accentuated by the use of ether extract of the uropygial gland of the loon, which exerts a powerful and specific olfactory stimulus. Low floating styrofoam decoys representing black ducks were used as the three-dimensional models, these being coated with Tanglefoot adhesive and provided with the attractant gland extract. The two-dimensional silhouette took the form of 'head−neck' decoys of Bristol board, as well as cylindrical models. These experiments showed that the visually attractive areas were at the angle of the head and neck, and at the prominent terminal portion of the model. Beyond that the pattern of fly distribution was influenced by the exact location of the scent bait.

7.3 Tsetse fly response to visual traps

New insight into the responses of biting flies to animal models and animal-model traps has been provided by observations on tsetse flies, part of a very extensive programme into tsetse behaviour in Zimbabwe (Rhodesia) spanning the 1970s and beyond. Reference to many aspects of that work will be made elsewhere in this review. In the design of experiments, and the interpretation of capture data, this tsetse orientated programme has yielded results which have relevance to many other fields of insect behaviour.

7.3.1 First experiments in Zimbabwe with animal models

The first series of experiments originated in the long established fact that *Glossina morsitans*, one of the principal tsetse fly species affecting wild game and domestic stock in East and Southern Africa, shows a feeding preference for certain game animals, the wart hog in particular, despite the relative scarcity of that host in most game populations. This problem was explored by means of an animal-model, roughly the same size as a wart hog, and conveniently provided by a 10-gallon black drum mounted horizontally on a small pram chassis about 14 in above the ground. Unlike other animal-models described so far, this had the additional advantage of being usable either as a static object or as a moving one. The model was moved along a straight 'run' at about 1.5 m.p.h. by means of a long handle held by an operator walking 23 ft ahead of the model, or alternatively by means of a long rope operated from a distance. The attractant for tsetse flies took various forms; either the model alone; model accompanied by walking man, or walking man alone. Tsetse visiting these attractants were counted approximately by the naked eye, or with the aid of a telescope. Tsetse settling on the traps were also retained for counting by means of a sticky deposit on trap surfaces (Vale, 1969).

The first and most striking and consistent observation was that more tsetse, *Glossina morsitans*, were attracted to the unaccompanied model than to the walking man. Even more unexpected was the observation that the model alone was much more attractive than the combination of model plus man walking 3 ft away. Using all possible combinations of experiment component, the results all pointed to certain conclusions of great significance to the whole problem of sampling tsetse flies. For this species of tsetse fly certainly, the presence of man close to the model attractant had a depressing effect on the catch, particularly of female flies. Further observations showed that the effect on the other main game tsetse species, *Glossina pallidipes*, was even more marked, affecting both sexes (both of which are blood feeders), and again especially the females.

This series of experiments, and others involving a wider range of attractants, all showed that the recovery of tsetse attracted to these models was best achieved by treating the model with a sticky adhesive, which proved to be markedly superior to catching by hand net. A further series of experiments was then carried out in which the black drum or 'model' previously used was treated with this sticky deposit, and a comparison made between (i) the drum alone, (ii) the combination with a man walking 3 ft away wearing a sticky screen, (iii) as in (ii) but the man not wearing a screen, and (iv) a non-sticky drum accompanied by a man who caught with a hand net any tsetse landing on the model. These were all operated along traverses of 700−1400 yds passing through a variety of tsetse woodland (Vale, 1974a,b,c).

In all these experiments, the unaccompanied sticky model trapped many more males and females, and a higher proportion of females, than any other combination. It was concluded that the low recovery by the hand net technique was due both to the depressing effect of the human catcher, and to the inefficiency of the hand-catching technique itself. The important finding is that with *Glossina morsitans*, and even more so with *G. pallidipes*, the presence of man — whether in the role of bait, catcher or observer — had a diminishing or even repellent effect on tsetse attraction to the model bait. No doubt this is associated with the fact that man is probably not the normal host of these species of tsetse. In the case of extremely hungry flies, mainly males, which have been unable to obtain a blood meal elsewhere, this repellence can be overcome, and the response is partial rather than absolute (Hargrove, 1976).

In view of the role of the human presence dictating these results, the experiments then extended logically to trials with a man-model, which was life size, painted a non-shiny black and clothed in a khaki uniform. Both live and model baits were fitted with sticky surfaces to trap alighting tsetse. In the upright stance, the model showed the same degree of non-attraction as a live upright man, indicating that visual stimuli play a dominant role in this type of repellence.

7.3.2 Experiments on the presence of man: electrified grids

These findings dictated that in a second series of experiments, the presence of man — whether as fly collector or as observer — had to be excluded. The tsetse flies attracted to such animal-models had to be trapped exclusively by mechanical methods, mainly by use of sticky surfaces (Vale, 1974 a,b). The sticky compound itself was critically tested in order to ensure that it did not possess some intrinsic attraction. The original oil drum model was replaced by a horizontal cylinder of metal or fibreglass, 50 cm long and 37 cm in diameter, completely covered with cloth and mounted 37 cm above the ground on a metal frame, the frame being provided with spoked wheels when used for mobile tests (Figure 7.2).

In addition to the use of sticky deposits, now coloured to match the baits, entirely new techniques were introduced based on the electrocution of flies making contact with electrified grids or surfaces. Run from 12 V batteries, the electrified grids of blackened fine steel wire running 0.8 cm apart could be set up over the attractant, for example on a wooden board on a man's back or used in conjunction with fine nylon netting to form an electric net to catch tsetse in flight. Electrocuted flies were recovered from trays coated with adhesive, or by other non-return devices. Both males and females are highly susceptible to electrocution, and the grids produced an initial knockdown of 96–99%.

At this stage, experiments with live bait animals were introduced into the series, and will be discussed elsewhere (Section 8.3.4). These observations endorsed the results of the work on animal models in showing that in mobile baits, visual stimuli play a dominant role in initially attracting tsetse flies, odour alone being insufficient to provide this stimulus. In the case of static models, or static bait, odour alone plays an important part in attracting tsetse to the vicinity of the bait, this orientation being

Figure 7.2 *Catching devices for tsetse flies. Man with electric net. Electrified decoy drum. Man with electric back-pack (Vale, 1974c).*

rendered more precise if additional visual stimuli are offered, particularly in the vicinity of the bait. In interpreting these findings, two factors have to be borne in mind. First, in the normal animal host stationary periods alternate with moving ones, exposing the same animal to a range of conditions covered by the animal-model tests. Secondly, tsetse flies themselves pass through a series of phases or hunger cycles, at each of which different stimuli may play different dominant roles in attraction to either bait or model.

7.3.3 Other visual traps used in tsetse fly survey

Before returning to the experimental programme in Zimbabwe on tsetse response to traps, it would be useful to see what lessons can be learned from the experiences with tsetse in many other African countries. The use of added attractants such as CO_2 and ox odour, combined with more effective methods for retaining trapped or attracted flies by means of sticky trapping surfaces and electrified grids, has all tended to show that while visual stimuli play an important part on close approach to the trap, it is the stimulus of odours which initially attracts most flies to static traps, as distinct from mobile ones, from a distance. The need for visual traps to bear resemblance to host animals has become diminished in importance, and most of these in current use can no longer be regarded as 'animal-models'. Provided the three-dimensional trap presents the necessary contrast between light and dark areas correctly positioned, the animal shape is not an essential ingredient.

In the constant search for new and more efficient mechanical traps, many new designs have been created, all based on the common principle of attracting flies to the inside of the trap in such a way that their subsequent movements in flight funnel them through a series of net cones and baffles into a non-return cage (Hargrove, 1977).

To the long list of traps named after their respective designers, the most recent trends and improvements are best exemplified by the vertical vane traps devised in East Africa and Zimbabwe (Hargrove, 1977) and the biconical trap designed by French workers in West Africa (Challier and Laveissiere, 1973; Laveissiere *et al.*, 1979) and subsequently adopted in Nigeria (Koch and Spielberger, 1979). Both traps, designed quite independently in separate regions of Africa, have several essential features in common. The main components of the vertical vane trap are (a) a model or visual bait to attract tsetse to the trap; (b) a condenser, or non-return device to restrict any flight away from the bait; (c) a collector or removable box to retain the flies in a compact volume; and (d) a 'hat' to concentrate the flies below the collector (Figure 7.3).

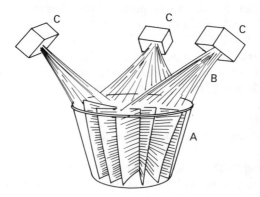

Figure 7.3 *Vertical vane trap for tsetse (Hargrove, 1977). A. V44 condenser. B. Hat, C. Collector.*

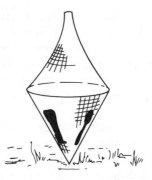

Figure 7.4 *Biconical trap for tsetse (Challier and Laveissiere, 1973).*

The biconical trap (Figure 7.4) is based on the same principle of attracting tsetse flies in flight to contrasting dark and light surfaces, directing their movements inside the trap to the daylight coming through an opening in the top, and retaining them by means of a non-return exit cage or collector. In this model, the dark shady openings are set in the side of the lower cone, and not underneath the trap as in many other models. Originally designed for the riverine tsetse of West Africa, this trap has been found to have wider use in other regions, and has been the subject of a critical examination with the East African game tsetse (Flint, 1985).

Experiments on its improved performance have dealt with a range of colour and shade combinations, both inside the trap and outside. For example, comparisons were made when the whole of the outside of the lower cone was white, black or blue, and also with the same range of colours on the inside of the lower cone. In addition, black on the upper cone and white on the lower cone was compared with white on the upper cone and black on the lower cone. When the original biconical trap was used for catching a sample of tsetse at dense population levels, it performed efficiently on its own, without any odour attractant. Now there is increasing use of chemical attractant to amplify the effect, particularly at low population densities. The attractant in Flint's studies (1985) acetone was released at the rate of 0.5 to 5 g h^{-1}, either on its own or through the medium of a CO_2 dispenser; this additional element can result in doubling the catch of tsetse.

The work in East Africa has also been concerned with further improving trap design for general application to the game tsetse, while at the same time bearing in mind such practical considerations as cost, ease of transport and ease of assembly. The improved design ultimately selected was called the F2 trap, and its construction and mode of action is amply illustrated in Figure 7.5 (Flint, 1985). This F2 trap was found to be ten times as effective — when used with acetone odour attractant — as the biconical trap for one of the principal game tsetse flies, *Glossina pallidipes*. It was also twice as effective for females of the other main species of game tsetse, *Glossina morsitans*, but the two traps showed equal performance with male *morsitans*.

All these experiments revealed another significant factor in visual attraction, the ageing of the trap. When traps are new the cloth is bright white, but after a week of exposure to the sun it becomes off-white, and after a month develops a distinct yellow−grey tint. By this time the catch may decline to only about 10% of the initial level. Confirmation that this is a visual effect is provided by the fact that repainting these surfaces to the original bright white restores the tsetse catch to normal.

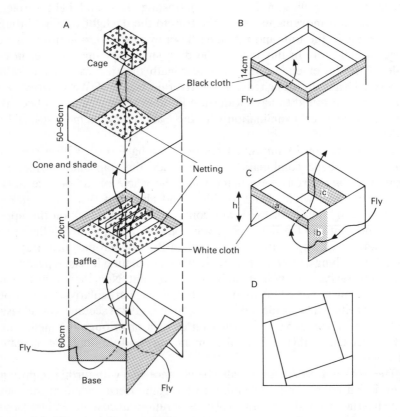

Figure 7.5 *F2 traps for tsetse (Flint, 1985).*

7.3.4 Flight patterns of tsetse approaching traps

In the continuing work in Zimbabwe on the design of more efficient traps for tsetse flies, two main objectives have had to be borne in mind. The first consideration is a trap which will attract and retain tsetse over a wide range of fly density, with particular emphasis on efficiency at very low densities at which some species of tsetse continue to be effective vectors of human and animal trypanosomiasis. The other consideration concerns the possibility of tsetse fly control by sterilization, and on the design of trap which will automatically capture, sterilize and subsequently release the treated flies back into the natural population. Whatever the objective, the same essential requirement is to examine all features of trap design which could lead to improved trap performance. For this purpose a much more critical analysis is required as to exactly how tsetse flies respond during the series of flight movements which ultimately direct them into the body of the trap (Vale, 1977a).

As mentioned above, tsetse are large robust insects easily visible in their day-time activities by naked eye or by telescope. However, any possibility of effective direct observation on their behaviour is ruled out, either because the human presence would distract flies from the trap, or — as the work in Zimbabwe has already disclosed — act as a repellent to flies initially attracted to the vicinity of the trap. Accordingly, in the next phase of research (Vale, 1982), the full potentialities of flight interception by means of electric grids, used so successfully in the experiments already described, have been further developed. This has provided new techniques, and new insight, into all aspects of tsetse flight response at all phases of attraction, entry and subsequent retention by the trap.

A variety of 'trap-like objects' were made with a frame of flat iron bars welded together, matt painted and covered with cloth and netting. All the cloth used was non-shiny cotton, and all netting was white nylon mosquito gauze. The effect of all traps was enhanced by a current of odour from 1−3 oxen in a ventilated pit. Flies in flight were electrocuted by contact with vertical sheets of fine black netting provided with a grid of fine black electrocuting wires. This system was found to kill about 90% of tsetse flies making contact. The electrocuted flies dropped into a tray of sticky adhesive for retention and examination.

The first objective was to determine the concentration of flies near the trap, and the abilities of various traps to bring flies close to their surfaces from a distance of several metres. For this purpose each trap was placed in a hexagonal cage of electrocuting netting 1.35 m high. This in turn was surrounded by four electrified nets arranged in an incomplete ring which covered about half of the arc round the trap (Figure 7.6A). Of the flies

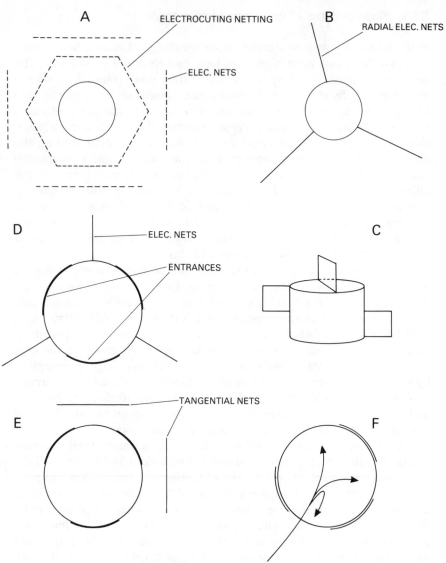

Figure 7.6 *Arrangement of tsetse traps and electrified netting (Vale, 1982). See text.*

that entered the ring, those that concentrated close to the trap were caught by the hexagonal cage. Of those which did not concentrate, some were caught on the inside surfaces of the incomplete ring of nets as they flew away from the traps. Of the total number of flies departing in this way, about half would be caught by the incomplete ring which covers 50% of the egress arc. At the end of the day, the total number of flies entering the ring is estimated by the catch on the hexagonal nets plus twice the catch on the incomplete ring. The traps tested took the form of various upright black cylinders on the ground, raised above the ground on stilts, cylinders which were all white, or were striped, etc.

Secondly, in order to study the distribution of flies within a few metres of the traps, the black cylinder on the ground was surrounded by radial electrocuting nets, at each of three positions from the trap (Figure 7.6B). In a modification of this arrangement, the question of heights at which tsetse fly around, alight on, and enter traps was tested by suspending the radial nets at various heights near a single trap (Figure 7.6C). In this series all tests were replicated using one trap which was black outside and one that was white outside.

To test the effect of wind direction, traps were provided with similar electrocuting nets on the upwind side and on the downwind side. Tests were also carried out on the effect of shape and size of entrance, with regard to square versus rectangular and horizontal versus vertical.

In order to investigate flight round the trap in relation to entry into the trap, the cylindrical trap was provided with three evenly-spaced entrances at the base, and three evenly-spaced electrocuting nets radiating from the trap surface (Figure 7.6D). With this competing arrangement of electrocuting devices, flies could be caught entering the trap only if their flight around the trap was very short. Few of the flies which flew straight towards the entrance would be caught by the radiating nets.

The arrangement of traps and nets for testing alighting and departing, is shown in Figure 7.6E. Some of the flies which approach within 1 m of the trap were caught on the outside of the nets. Of the flies which moved closer to the traps, those which entered the trap were caught by an electric net, those which alighted were caught by the surface grid, and some of those which departed were caught on the inside of the tangential nets. Of the flies which visit the trap without alighting, only about one-quarter of *G. morsitans* and one-seventh of *G. pallidipes* were found to enter. It was also shown that flies which exhibit alighting reactions could, subsequently, be much more likely to enter than flies showing no alighting reaction.

Superimposed on this basic experimental layout, tests were also incorporated on the effect of various surfaces, such as dull black, dull white and shiny black, and also with one-half of the trap white and the other black, etc.

Further tests also examined the effect of entry size, bearing in mind that entry can also provide an exit for the flies. A black cylinder 1 × 1 m with white interior and netting top was provided with three evenly-spaced entrances at the base: a 30 × 30 cm entrance to admit flies; a 30 × 30 cm entrance covered with a netting grid to capture flies flying out of the trap; and a 50 × 50 cm entrance with a similar grid (Figure 7.6F). It was found that the number of flies leaving through the 50 × 50 cm entrance was 1.84 times the number leaving through the 30 × 30 cm entrance. Thus, large entrances encourage exit response to roughly the same degree as they encourage entry.

7.3.5 Development of methods to improve and supplement biconical traps

The biconical trap is now a firmly established visual trap for tsetse flies in several West African countries where it is efficient in capturing both the savanna species *Glossina morsitans* and the riverine *G. palpalis*. However, research workers are continually aware of the need to develop alternative sampling or trapping techniques which can be used concurrently, and help to build up a more accurate picture of the ebb and flow of tsetse populations under all environmental conditions. Two of these alternatives have been explored in detail.

(a)   The water trap
This is a comparatively new principle introduced into tsetse studies (Deansfield *et al.*, 1982). These traps consist of enamel trays 31 × 26 × 5 cm filled with water to within 0.5 cm of the top. To each trap is added 10 ml of detergent and 5 ml formalin. Traps were placed on stands 40 cm above ground level. Over the water trap, different arrangements of upright baffles made of polyvinyl chloride in cross form were set up, and these could be painted black or white. The range of colours available for both traps and baffles enabled a wide variety of combinations involving black, dark blue, yellow and white to be tested. The experiments also included the effect of presence or absence of detergent in the water.

The performance of these different trap combinations was monitored by means of concurrent captures in static biconical traps for comparison of the two savanna species *G. morsitans* and *G. tachinoides*. These tests showed that the water trap acted visually as a horizontal colour screen, in which the intensity of reflected light — particularly in the ultraviolet — could be a key factor. White was found to be the most attractive colour for *G. morsitans*, and blue for *G. tachinoides*. The catch in the white water trap was increased by black baffling.

(b)   Visual response to two-dimensional coloured screens and to three-dimensional traps

In the continual search for new and more effective methods for trapping and controlling tsetse flies, ever increasing attention is being given to the role of visual attraction which will guide tsetse to certain surfaces in preference to others. This information plays a vital part in control strategies which depend on attracting tsetse fly, with or without additional olfactory lures, to insecticide impregnated surfaces. The visual role of solid or three-dimensional objects or traps has already been discussed. The outcome of that work leads naturally not only to a more critical examination of attraction to screens, but also to the question of the part played by colour in visual attraction.

Most of the basic work on visual attraction of tsetse to traps involved experiments with various combinations of black and white, with less attention being paid to the colour factor. In the cube-shaped F2 model trap developed in Zimbabwe (Flint, 1985) the standard colour of the cloth was white, with black interior portions. In a series of experiments, the white cover was replaced with a series of coloured materials, and it was found that those which selectively reflected in the blue−green and red bands tended to be attractive, while those in the green−yellow− orange and UV bands were unattractive (Green and Flint, 1986). Bright royal blue emerged as the best trap colour for *Glossina morsitans* and *G. pallidipes*.

When extending these experiments to screens or panels, as distinct from established three-dimensional biconical or F2 cube traps, it became essential to devise some method of retaining, or trapping, the flies attracted to flat surfaces. This was achieved by adding panels of mosquito netting at each side of the 1 m^2 test panels. These act by intercepting flies which are attracted to the panel, and circle around it. When these side nets are electrified, tsetse approaching close to the surface of the screen are stunned or killed.

In a series of experiments with *Glossina palpalis* (Green, 1989) it was found that − like *G. morsitans* and *G. pallidipes* in Zimbabwe − bright blue was strongly attractive to these tsetse. A particular colour, phthalogen blue, was found to be highly attractive, but did not induce a strong landing response. Other hues, reflecting strongly in the ultraviolet, induced landing responses but were not initially highly attractive to the flies. The idea of combining these two complementary colour qualities led to experiments with bi-coloured screens as an alternative to single-coloured screens.

The 1 m² screens or panels were tested either on their own, or flanked by mosquito netting panels which stunned or killed approaching tsetse. The two-coloured screens were made of strips of equal area (Figure 7.7). Colours which induced a strong landing response were compared with those which were strongly attractive, both as single-coloured screens and as bi-coloured ones. In all captures, separate records were made of male and female captures.

The combined phthalate blue and white (UV reflecting) screen was found to be the most efficient in capturing females, and was moderately successful with males as well. Various alternatives in the way of screens divided vertically, horizontally and diagonally were also tested (Figure 7.8). These showed that the best arrangement of the bi-coloured trap was the diagonal one, with the superior triangle phthalate blue, and the inferior triangle UV-reflecting white. On a simple screen, with no side netting panels, this combination caught 2.4 times as many flies as an all-blue screen. The addition of side or flanking panels of mosquito netting increased the catch four times in an all-blue panel, but in blue and white screens, netting panels only increased the catch by a factor of two for females and three for males. Evidently, colour combinations are useful for simple screens, but not for flanked ones. The interpretation of this is that in simple screens, both of the distinct responses — attraction and landing —

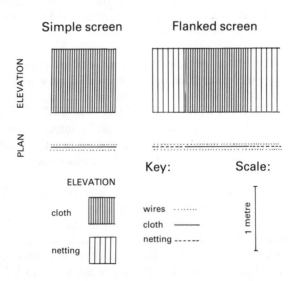

Figure 7.7 *Design of simple screens and flanked screens enclosed in electric nets used in tsetse fly investigations (Green, 1989).*

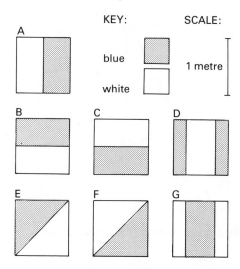

Figure 7.8 *Design of blue and white bicoloured screens used in tsetse experiments (Green, 1989).*

affect the final catch, but for the flanked screens, only the attractant element operates.

7.4 The Manitoba horse-fly trap, three-dimensional silhouette

The Manitoba trap, first designed and tested in the Canadian province of that name, is essentially a plastic cone resting on a tripod, with a collector/container fitted at the apex (Figure 7.9). Suspended beneath the cone is the attractant object in the form of a black sphere (Bracken *et al.*, 1962; Thorsteinson *et al.*, 1965). The concept of this trap design originated in the observation that horse flies were attracted to a weather balloon. A balloon of this type was then tested by inflating it to 24 in diameter, coating it with Tanglefoot, and suspending it 4 ft above the ground. Later, spheres which could be coloured were suspended as a decoy below a fly trap with a translucent canopy. The rubber balloons were later replaced by polystyrene spheres inflated to 20 in diameter, and the non-return trap was charged with sodium cyanide in order to kill flies immediately on entry. In its final design, the Manitoba trap consisted of a spherical decoy target made from two hemispheres constructed of black acrylic plastic sheets.

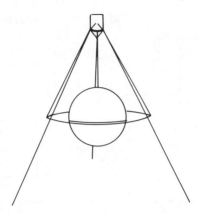

Figure 7.9 *Manitoba horse-fly trap (Thompson, 1969).*

The trap chamber was capacious enough to hold approximately 2000 horse flies, and the trap performance can be judged by the fact that in some areas capture rates of the order of 1000 female tabanids per hour were recorded. Trap capture data have formed the basis of information about seasonal and geographical distribution of Tabanidae in Canada, as well as essential data on species composition (Hanec and Bracken, 1964).

7.4.1 Experiments with original, purely visual, design for tabanids

The potentialities of the Manitoba trap have been fully explored by American workers in connection with the horse flies (*Tabanus* spp) and deer flies (*Chrysops* spp) which are widespread biting pests of domestic stock over extensive areas of eastern North America. One of these species, the saltmarsh greenhead, *Tabanus quinquevittatus*, has been the prime target for study.

In one direction, studies on the Manitoba trap — which is attractive to females only — have been concerned with comparisons with other trapping techniques such as the Malaise trap (Section 5.2), which captures both males and females, with older methods of hand-collecting flies on live ox bait, and with simulated animal traps such as the Morris trap (Section 7.1) (Thompson, 1969; Thompson and Pechuman, 1970; Thompson & Bregg, 1974). Other American studies have been concerned with modifications and improvements of the basic design of trap itself. In one of these studies (Hansens *et al.*, 1971) a comparison was made firstly with the black sticky panels which had been a standard trapping method for many years before the development of the Manitoba trap. This was followed by experiments on variations of the black decoy sphere in height and position, and increases in the decoy size by means of a black plastic 'skirt'. Tests were also carried out with substituting a black fibreboard barrel for the standard black sphere.

All of these trials confirmed the efficiency of the Manitoba trap, which was capable of trapping several hundred flies per day. It was also found that increases in catch were produced by raising the barrel decoy within the trap to an optimum height of 20 in, and also by the addition of the black plastic skirt. In all of these modifications, catch was still further increased by the addition of CO_2. The experiments involving sticky panels also pointed to the importance of areas of contrast as an important factor in attracting flies to the Manitoba trap itself, a factor which is dealt with in more detail on page 221. In all the various test areas in Canada and the USA, the Manitoba trap was revealed as a valuable trapping method, being capable of sampling all known species of *Tabanus*, and being highly selective for this group of insects as against others.

These experiences have by no means been confined to America. In more recent studies on sampling tabanid pests of horses in the Camargue district of southern France (Hughes *et al.*, 1981) capture records showed a general correlation with hand-netting them as they attack horses. The model of trap used conformed closely to the original design, the attractive component being a shiny black ball, 400 mm in diameter, with the usual transparent cone above leading into a non-return cage.

7.4.2 Experiments on headflies (*Hydrotaea*) in the UK

The most recent and penetrating study of insect response to Manitoba traps has been carried out, not with tabanids, but with a muscid fly, a pest species of cattle in parts of Great Britain. The target species in question is the headfly, *Hydrotaea irritans*, which is not strictly a biting or blood-sucking insect, but feeds on exudates, secretions and oozing wounds and sores on cattle. Nevertheless, the visual attraction to the animal host plays a dominant part in its behaviour as with other true biting flies active by day.

(a) Experiments with original model
These experimental studies, which were carried out by two separate groups or teams, can be conveniently divided into three stages or progressions. In the first phase (Berlyn, 1978a,b) the design of Manitoba trap was strictly in accord with the original, and comprised a black sphere 0.5 m in diameter suspended beneath a polyethylene collecting cone, which was supported by a tripod. Flies attracted to the sphere flew

upwards to escape, and were trapped in the container at the apex. The trap catch was compared with the catch obtained by alternative trapping techniques running concurrently, principally suction traps with and without added CO_2. These comparisons showed that the Manitoba trap caught more flies than suction traps alone, but less than suction traps with added CO_2. The catches in the Manitoba trap confirmed the visual stimulus of the decoy, but also left open the extent to which the insect reactions were affected by thermal stimuli created by the heat absorbed by the black sphere. Both males and female flies were captured, with females predominating.

It had been hoped that the unbaited suction trap would provide a valuable non-selective baseline for sampling the entire cross-section of the *Hydrotaea* population. However, it was found that the suction trap catch made in the normal way collected far more adults than when the trap was camouflaged with bracken fronds, from which it was concluded that the visible black casing of the trap attracted flies in the same way as in the Manitoba trap.

(b)   Experiments with modified design, incorporating CO_2

A further stage in progress in the experimental studies involved modifications in the original Manitoba model (Berlyn, 1978c) in such a way that attractants could be hung beneath the polythene cone or canopy. A tube attached to the open apex of the cone also allowed CO_2 to be introduced. The 0.5 m attractant spheres were presented in six different colours, matt black, shiny black, shiny red, shiny green, shiny yellow and shiny white. A seventh trap without a sphere provided the control. In addition to the main target of research, *Hydrotaea irritans*, these traps also attracted large numbers of *Haematopota pluvialis* a biting tabanid dominant in those parts of Britain. The tests on the olfactory stimulant CO_2 were carried out concurrently using three traps with suspended red spheres, from the middle one of which CO_2 could be released at a range of concentrations from 0.5 to 5.0 $1 \, min^{-1}$.

In a further series of tests, potential attractants were suspended in turn below the canopy of the trap. These were (a) heat, provided by a hot fan heater, (b) a moving red sphere in contrast to the normal static one, as well as several potential odour attractants and repellents.

Matt black, shiny black and red attracted most flies to the Manitoba trap, and revealed a sexual distinction in that, unlike females, males were significantly attracted more to matt black than to shiny black. *Haematopota* was also mainly attracted to black spheres, and it was observed that on landing, these flies were surrounded immediately by individuals of *Hydrotaea* which thrust their heads close to the tabanid's proboscis, presumably to ingest any blood exuding from the wound made by the insertion of the latter's proboscis.

The interactions of colour and carbon dioxide suggested that the latter is of primary importance at the start of the search for the bait, but that visual stimuli — provided in this case by the decoy of the Manitoba trap — is important in the final approach. This would explain why traps baited with CO_2 only were observed to catch fewer flies than an unbaited red one only 2 m away; most flies evidently required the final visible stimulus provided by the red sphere. Of the other factors influencing attraction to the Manitoba trap, heat was shown to play a significant role. This element was examined by means of further modifications to the Manitoba trap carried out by a different research group (Ball and Luff, 1981) which leads to an arbitrary third stage in progress.

(c) Effect of trap temperature
In this stage, study was concentrated on three factors affecting trap performance and mode of action, surface temperature of the visual target, CO_2 and its interaction with temperature, and target size. The design of the modified Manitoba trap is shown in Figure 7.10. In place of the conventional black sphere as target, a 5-gallon drum was used, which was painted matt black and could be filled with water which could be heated and controlled thermostatically. Other drums contained unheated water at ambient temperature; CO_2 was supplied to both hot drums and cold drums.

The total number of flies caught on the heated drum was 1.7 times the number on the cold drum. Very similar results were obtained when the heating differential was provided by a fan heater releasing hot air under the trap (Berlyn, 1978c), suggesting that it is not the surface temperature of the target which is important, but the flow of warm air from the trap.

It appeared that a Manitoba trap with a warm target and CO_2 supply could mimic many features of a live target animal, and that the classification of Manitoba traps as mainly visual decoys may be an oversimplification of a complex situation involving several different factors.

In order to test the working hypothesis that the modified Manitoba trap appeared simply as a simulated bait animal (cow) to various Diptera associated with cattle, trap catches were compared with samples of flies caught by sweeping a hand net around the head and flanks of cattle (Ball, 1983). In addition, a 2×1 m electrocuting grid, placed downwind of a CO_2-supplied drum, was used to intercept flies travelling to the artificial bait. In this way it was hoped to obtain some measure of the actual efficiency of the trap. A complicating factor in what might appear to be a valid controlled comparison was that many species of fly netted around

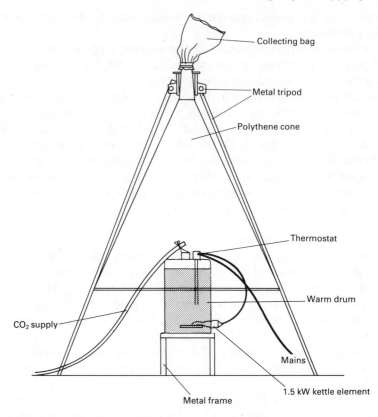

Figure 7.10 *Manitoba trap provided with heated target and CO$_2$ supply (Ball and Luff, 1981).*

cattle were not those primarily attracted to the animal itself, but those attracted to cattle dung, or disturbed from vegetation by the movements of the human catchers. In addition, the behaviour of the cattle — whether they were lying down or standing up, and whether they were isolated or part of a herd — all introduced additional variables affecting fly catch which were not exhibited by the static Manitoba trap.

Allowing for these factors, one constant difference was shown by the three most abundant species associated directly with cattle, *Hydrotaea*, *Haematopota* and the muscid *Morellia*, and that was that the proportion of males of all three were significantly higher in the samples from the Manitoba trap than on the cattle. All of these observations confirmed that for true cattle-visiting species of fly, as exemplified by *Hydrotaea*, Manitoba traps were again confirmed as being highly selective, attracting few non-cattle-visiting species.

7.5 Impact of background on visual traps

A great deal of attention has been paid to the composition, appearance, shape and colour of visual traps themselves, but rather less is known about the visual impact on the insect of the contrast between trap surface and background in the field. This subject has been rather more critically examined in the laboratory (Brady and Shereni, 1988), and although much of that experimental work is not directly concerned with responses to traps *per se*, the analysis of insect behaviour under those conditions has considerable relevance to the interpretation of observations made in the field.

One insect which has received especial attention from this aspect is the biting horse fly, *Tabanus nigrovittatus*, in the USA (Allan and Stoffolano, 1986). As part of their day-time host-seeking behaviour, tabanids are attracted to various visual objects, of which the Manitoba trap has already been described. They are also attracted to panels, which, when treated with adhesive, form sticky traps. The attraction to such traps is entirely visual, with no odour response involved.

Panels of this kind, which selectively attract mated uniparous females, were used in this experiment. The traps consisted of a background panel 60.8 × 60.8 cm, and test panels 30 × 30 cm which could be painted with different colours and covered with adhesive. In order to study first of all the effect of intensity contrast, i.e. the amount of incident light reflected, without the interaction of hue, the dominant wavelength of the reflected light, grey background panels were used. Three intensities of background grey panels were tested: (i) equal to the background vegetation, (ii) higher than background vegetation, and (iii) lower than background. Grids over the panels allowed the number of flies attracted to each separate square to be counted, and a distribution pattern determined. These counts were supplemented by close range observation on fly behaviour by means of binoculars.

The experiments showed that the attraction of flies to grey test panels was greatest when the contrast between test and background was highest. Maximum numbers were attracted to high intensity blue panels against low intensity grey background.

The distribution of flies on test panels was found to be highly clumped, as also was the case on the background panels, and top and bottom halves of the entire trap. The largest number of flies was collected along the boundary of the test panel and the background panel, but there was not great attraction to the edges of the background panel. The attraction was predominantly to the contrast between test panel and background panel, and not to the contrast between the background panel and the background vegetation.

The largest accumulation of flies occurred at a height of 22.8–30 cm above ground, a response in accordance with the fact that the flight level of cruising *T. nigrovittatus* is slightly above the level of the salt marsh grass characteristic of its natural environment. The response therefore shows not only a distinct edge effect, but also a preferred height of landing. The close visual watch showed that of the total flies observed, 20% landed immediately, 40% of the flies flew past the target, 31% flew past but returned and landed, while 8% flew past, returned but did not land.

7.6 Visual responses of the stable fly, *Stomoxys calcitrans*, and allied muscids to traps

Of the various trap components which determine visual attraction, the role of light-reflecting surfaces is well illustrated by American experience with the biting stable fly, *Stomoxys*, and its non blood-sucking allies *Musca domestica* and *M. autumnalis*, all of which are typical farmyard flies. For many years simple box traps were used for trapping stable flies, containing an inverted cone inside the box; but in the early 1970s, a discovery was made which completely dictated the future design of fly traps. This was the finding of the great attraction exerted on flies by the translucent fibreglass Alsynite (Williams, 1974). Panels 35 × 45 cm were fitted together with intersecting planes with the vertical flat surfaces fitted onto a wooden stake about 130 cm above ground. The surfaces were covered with the sticky adhesive Tack-Trap. In trials these traps were found to be seven times more effective than box traps, and this advantage was even more marked at low population densities. In a further examination of the peculiarly attractive quality of clear Alsynite it was found to be the most attractive of 30 other surfaces tested in the field (Agee and Patterson, 1983), and was characterized by an increased spectral reflectance in the 380–420 nm wavelengths, traps working most effectively when exposed to full sunlight which enhanced the reflection of the UV light. Clear Alsynite is highly attractive to both sexes of *Stomoxys*, and to flies of mixed ages.

The strong attraction to Alsynite was not shared by the allied non-blood-sucking muscid, the face fly *Musca autumnalis*, which was more attracted to the black and white boards which formed the basis of previous fly traps designed for general use. A combination of black and white pyramid traps for *Musca*, and Alsynite traps for *Stomoxys*, was found to provide the best practical solution.

Alsynite continues to be the essential basis of visual traps for *Stomoxys*, although the actual dimensions of the panels tend to differ with different workers. One unusual modification has been the subject of critical tests; in these the Alsynite surface took the form of a cylinder in contrast to the two-panel design used up till that time, the cylinder being made from a piece of fibreglass 90 × 30 cm in diameter (Broce, 1988). With this design, some factors were examined in more detail such as the thickness of plastic — using 1, 2 or 3 layers of Alsynite — trap size, traps of equal height but different diameters, 15.0, 22.5, 30.0 and 37.5 cm, and also performance in relation to wind direction. These experiments showed that the number of stable flies caught and recorded as 'flies per unit of adhesive-coated surface area' was directly related to trap diameter, but that in general there was no obvious difference in performance between the cylindrical trap and the more usual four-winged design. The house fly, *Musca domestica*, was also found to be attracted to Alsynite, but to a lesser extent than *Stomoxys*. The efficiency of the adhesive surface itself tends to deteriorate quickly due to build up of dead flies, dust and debris. Even after only 24 h exposure the catch can fall off to only 40% of its original figure (Agee and Patterson, 1983).

7.7 Mosquito reaction to visual targets and traps

A great deal of work has been done on the responses of mosquitoes to coloured visual stimuli, and this has shown that different species, even of the same genus, can respond differently to the same visual stimulus (Browne and Bennett, 1981). However, only in rare cases have these visual reactions been used to design an actual trap, as distinct from an attractant panel, and much of that information does not come strictly within the theme of 'trap response'. The logical idea of coating attractive panels with adhesive, thus converting them into traps — so effective with tabanids and stable flies — was found to be quite unsuited to day-flying and crepuscular mosquitoes in Canada. Mosquitoes were observed to hover above the surface of the experimental targets, and extending a leg to 'feel' the surface before landing. On contact with the sticky surface, the mosquito could reverse flight with sufficient power to pull one or two legs free from the adhesive coating.

The problem of actually designing a trap in order to study colour response has only been tackled experimentally, but much can be learned from the design and operation of these. With the failure of two-dimensional sticky traps, attention was concentrated on three-dimensional forms, initially six-sided cuboid measuring $254\,cm^2$ per side. Five faces of each cube were covered with coloured art board, in each of which was centred a small plastic funnel providing access for the mosquito to the centre of the cube through the narrow, 13 mm end of the funnel. The sixth face was provided with a mesh sleeve. Five colours of art board were used, black, blue, red, yellow and white.

Mosquito responses to three-dimensional targets of different shapes were also studied — cube versus pyramid — as well as the responses to different regions of a two-dimensional rectangular target. The results obtained with the day-time biting mosquito species, *Aedes cantans* and *A. punctor*, and the crepuscular *Mansonia perturbans* showed overall a prime attraction to black, followed by red and blue to about an equal extent, with white and yellow being quite unattractive. Faced with a choice of three-dimensional shapes, cuboid and pyramid, *Aedes cantans* and *Mansonia perturbans* found the cube about twice as attractive as the pyramid even though both targets had approximately the same surface area. These two species also showed strong preference for the projecting parts of a rectangular target. It should be noted that all these visual targets were provided with an added stimulant or attractant in the form of CO_2 in order to guide mosquitoes flying in the vicinity of the traps, at which point visual reactions to colour and shape become dominant.

Response of mosquitoes, both by day and by night, to visual stimuli associated with traps has already been revealed as a new and complicating factor in the action of traps which are supposedly non-attractive, such as flight or interceptor traps, as well as suction traps (Section 2.4). The interpretation of capture data must allow for known, or unknown, visual responses by the mosquito.

Chapter 8

Animal-baited Traps and Animal Odours

8.1 Introduction

The winged blood-sucking insects of main concern here all belong to
the order Diptera (two-winged flies) and include mosquitoes, blackflies
(*Simulium*), sandflies (*Phlebotomus*) biting midges (*Culicoides*) tsetse
flies (*Glossina*), horseflies (*Tabanus*) and stable flies (*Stomoxys*). A very
wide range of trapping devices for these insects have evolved over the
years (Service, 1976, 1977a,b; Muirhead-Thomson, 1982). Those trapping
systems which are based on attraction to light, on response to visual

225

targets, and on flight interception by means of flight traps or suction traps have already been discussed. All of these groups of insects have the common factor of a blood-feeding habit, either on man, his domestic animals, or on wild, feral mammals and birds. This dependence on the animal host for regular blood meals by the female blood sucking fly, and by both male and female tsetse flies, has provided a basic method of capturing all these insects, whether they are predominantly day-biting such as tabanids, tsetse, blackflies and many *Aedes* mosquitoes, or mainly nocturnal as with the majority of mosquitoes, midges and sandflies.

In this simplest form, captures of blood-sucking flies attracted to the host animal are carried out by collecting the insects — by aspirator — as soon as they settle on the host. The host may be either man, in which case the collector normally captures insects settling on his legs, or on the body of a human bait, or on a tethered animal such as a calf or ox. Collections on human bait still play an important role in many biting fly studies, and in fact form the mainstay of vector monitoring in the WHO onchocerciasis control programme in West Africa where the bulk of knowledge about the blackfly vectors involved — the *Simulium damnosum* group — is based on standardized human bait captures, recorded as flies per boy h^{-1}.

Somewhat similar techniques, involving a hand net rather than an aspirator, were for long standard methods of capturing tsetse flies in the 'fly round', and are still used from time to time as a control for comparison with new and experimental capture methods. Variations of this method include swinging a net around a host animal to capture attracted flies, a method still practised with the strong flying tabanids and others. None of these capture methods — which have all made major contributions to knowledge of behaviour and of population fluctuations of winged insects of medical and veterinary importance — involve traps as such, and therefore fall outside the scope of this review. This stricture also applies to the collecting of resting populations of blood-sucking flies, whether inside houses and animal shelters, or outside on vegetation or other shaded sites, and which in most cases do not involve traps.

In the traditional method of catching insects attracted to animal bait it was long appreciated that the presence of the human collector or observer was an undesirable factor likely to interfere with the correct interpretation of capture data. The main concern was that the human presence would act as a counter-attractant, amply confirmed in many cases by finding human blood in flies captured off animal bait. What was not appreciated until much later was that with some insects the human presence could act as a repellent to attracted flies, or as a deterrent to biting the bait animal in question.

8.2 The drop net or falling cage

8.2.1 Earlier work with tsetse flies

One of the earliest measures to eliminate the human presence in studies on the attraction of various animal hosts to blood-sucking flies was the drop net or falling cage. As used in early studies on tsetse flies in Africa (Phelps, 1968) a large net 9 × 6 × 7 ft was used which could be raised on a pulley and then quickly lowered over a tethered ox, allowing access at that point for a collector to enter the cage and collect the trapped tsetse. The practice was to lower the cage as soon as the animal arrived on site, and thus capture flies attracted to the moving bait. The cage was left down for 10 min to allow flies to be collected, and then raised again for 10 min to expose the animal while the human collector remained some distance away, and this continued throughout the day. Observations on this cage sample showed that the composition of the first falling-net catch of the day, made immediately at the end of the ox bait's walk to the station site, differed from that recorded on the static animal for the rest of the day, in that a smaller proportion of female tsetse were caught, but a higher proportion of these had blood fed. It was concluded that in the case of the two game tsetse species involved, *Glossina morsitans* and *G. pallidipes*, a moving bait animal is particularly attractive to hungry flies, while females are attracted to stationary ox bait for reasons other than for immediate feeding. The experimental follow-up of these observations has already been described (Section 7.3).

8.2.2 Earlier work on blackflies (*Simulium*) in Canada and Africa

Essentially the same principle has been used to trap blackflies (*Simulium*) attracted to mammal and avian hosts exposed in open mesh cages (Anderson and de Foliart, 1961). At the end of an exposure period of 15 or 30 min, a cardboard trap in the form of a 'black-out' box, slightly larger than the cage, was placed over the host cage. A panel in the black-out box could then be opened allowing captured flies, attracted by the daylight, access to a netting compartment where they were retained. The technique was found to be particularly useful in comparing the reactions of *Simulium* in the presence of alternative hosts, such as different species of bird, and could also be used, by means of a pulley system, to study the attraction of insects to the trap at different levels above ground.

A modification of this same principle has also been tested with the *Simulium damnosum* complex in Africa. Very little is known about the alternative hosts for this group, apart from its obvious attraction to man, but avian hosts would seem to be a likely choice. In order to test this experimentally, in the presence or absence of man, a cage was lowered periodically over the bait bird — chicken — (Figure 8.1). The host was freely exposed, so there was no question of mechanical or visual obstruction to the flight of approaching flies. By repeatedly raising and lowering the cage within each hour, and extending observations throughout the day, patterns of biting activity could be established (Disney, 1972). An essentially similar principle, but based more on the North American box design, has also been tested successfully with *Simulium* species in other parts of Africa (El Bashir *et al.*, 1976).

8.2.3   Experiments with biting flies in Australia

The drop net or falling cage has also been found particularly useful in Australia for sampling biting flies, especially those attracted to sheep tethered in a portable crush (Muller and Murray, 1977). In this case a pyramidal net was used, 1.8 × 1.2 m at base, and approximately 2 m high at apex. When the net is lowered, the human observer enters the cage through a sleeve in the side, and collects the insects by means of an aspirator.

8.2.4   Recent work on trapping biting flies in Canada

The design and operation of traps based on this principle have been examined critically by Canadian workers (McCreadie *et al.*, 1984) who have pointed out that for some biting insects, blackflies in particular, the presence of such a large visual object over the host may deter some

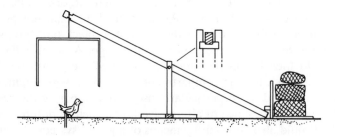

Figure 8.1   *Cage for lowering periodically over host animal (chicken) attractive to blackflies,* Simulium *(Disney, 1972).*

species. An alternative method in which a folded net on the ground could be remotely flipped over the host provided an alternative in the case of sheep (Jones, 1961), but it was considered that in order to utilize this method for larger animals, such as the cow, the sudden spring-trigger movement, and enclosure, would disturb the animal too much. The method devised was to set up a metal frame consisting of a rectangular base with three hoops spanning the width of the rectangle, which was fixed over the wooden animal pen. A collapsible tent of fine-mesh netting was attached to the side of the metal base. By means of rings sewn onto the net, the collapsible net could be quickly moved on the hoops over the bait animal to enclose it completely. An operator at each end guided the net by means of two rods, and the whole operation was completed in three seconds. A zipper running down the middle of one end allowed access to the inside of the tent for the collection of trapped insects by hand vacuum. The normal exposure period was 10 min, during which the human operator moved about 20 m distant.

This trap was successful in collecting 26 species of biting fly belonging to four families (mosquitoes, tabanids, biting midges and blackflies). In the absence of bait there was no evidence that the framework of the cage provided a visual attractant, at least with *Simulium*, nor was there any evidence that the presence of the trap inhibited blood feeding, judging from the high proportion of engorged females trapped.

8.3 Automatic trapping of insects attracted to live bait

8.3.1 Introduction to wide range of devices

Traps which mechanically retain biting insects attracted to live bait have taken many forms over the years, according to the species involved, the objectives of trapping and the ingenuity of the particular research worker. Many of these consist basically of large box structures with entry louvres or slits in the lower half, and a netting-covered top or cage into which insects attempting to leave the trap are directed by the light coming through from outside. The tethered animal bait — calf in many cases — may be protected from attacking flies by netting, or it may be exposed inside the trap, allowing entering insects to bite or feed. In studies on man-biting species, the bait is human in many cases, sleeping inside a bed net inside a larger net fitted with entry louvres, or suspended in such a way as to leave a narrow gap between the bottom of the net and the ground, allowing access to mosquitoes and other biting flies.

8.3.2 Recent critical assessment of the Magoon trap

Many of these trapping devices have been mainly concerned with determining the species and numbers of night-biting insects, commonly mosquitoes, attracted to different kinds of bait, and the actual problem of
trap response has not always received due critical attention. This certainly
applies to a calf or donkey baited design — the Magoon trap — long
popular in the USA and Central America to monitor populations of
swamp-breeding mosquitoes such as *Anopheles quadrimaculatus* and
A. albimanus in which very high mosquito populations are common.
Comparatively recently, (Muller *et al.*, 1981) this design has been more
critically tested by Australian workers studying blood-sucking mosquitoes
and midges in the northern tropical areas of that continent, in relation to
arbovirus transmission.

The Magoon traps were baited with either domestic chicken or calf,
enclosed in mosquito mesh to prevent bait animals from actually being
bitten. Despite the protection of the bait from biting flies entering the
trap, many blood-fed mosquitoes were collected, all of which were pre-
cipitin-tested to determine the source of their blood meals. The 16% of
these bloods which were found to be positive for man could be attributed
to ravenous mosquitoes feeding on the human collector when he removed
trapped mosquitoes from inside the trap by means of a motorized aspirator.
Some of the positive calf bloods could have been caused by accidental or
occasional resting against the netting by the calf. However, the considerable
number which had fed on ox blood, the 10% of insects from the calf-
baited trap which had fed on neither ox nor man, but on buffalo, horse or
marsupial, plus the 32% of blood meals from chicken-baited traps which
were not avian nor man, all indicated that the Magoon trap was acting as
a directional, visually-stimulating object, or that engorged mosquitoes
from outside were using these traps as resting places.

Further limitations of the traditional Magoon type of trap emerged
from studies on the mosquito vectors associated with severe epizootics of
western equine encephalitis in Argentina in 1982–83 (Mitchell *et al.*,
1985a,b). This trap was found to be quite inadequate for such larger
animals as horses used as bait in these studies. A large net trap, of a type
which had been used in many other parts of the world, was found to be
more suitable. In this case the net was 3.6 m wide, 3.6 m long and
2.1 m high. The base of the trap could be raised 30–46 cm above the
ground while the bait animal was inside. The net was lowered in the
morning before the trapped insects were collected from inside by means
of a mechanical aspirator. Over 1000 mosquitoes have been trapped in a
similar net cage overnight, and in this particular investigation 46% of
these were species of *Culex*.

8.3.3 Baited traps utilizing suction fans

A further range of trapping techniques for the retention of biting insects attracted to live bait involved the use of suction fans. This baited suction-trap, whose use is quite distinct from that of the suction trap operated alone (Chapter 2), usually takes the form shown in Figure 8.2, insects entering the lower part of the trap being sucked up into a collecting sleeve by the updraft of the fan blades at the top of the trap.

This technique has proved particularly useful with small avian or mammalian (rodent) bait animals for forest species of mosquito which normally feed on such hosts. A good example is provided by arbovirus research in Trinidad (Davies 1975, 1978). Virus isolates from two of these *Culex portesi* and *Culex taeniopus* suggested that the two mosquito species might have different host preferences. As most of the suspect host animals concerned were smaller than a rabbit, a standard sized baited suction trap could be used in which the propellor component could be activated by a time switch in such a way that it operated the fans for 45 s every $7\frac{1}{2}$ min. This interval allowed mosquitoes to approach the bait cage suspended below the trap, and for this number to build up for $7\frac{1}{2}$ min, at which time the air current produced by the propellor sucked the mosquitoes into the trap, egress from which was prevented by a valve. A series of similar traps arranged around the circumference of a circle was used to compare the attractiveness of 11 species of rodent, three marsupials, two birds, a bat, a reptile and a crustacean.

The baited suction trap has also proved valuable for sampling mosquitoes or other biting insects attracted to monkeys involved in the forest virus cycle. Experience from other regions had confirmed that when such a host is retained in a cage in the field, the structure of the cage and the

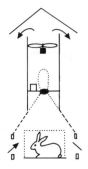

Figure 8.2 *Baited suction trap for mosquitoes and other small biting flies (Bidlingmayer, 1974).*

unavoidable mesh or netting, act as a deterrent to attracted mosquitoes. This was considered the main reason why many species of mosquito, which feed readily on such hosts in the laboratory, failed to be recorded in such traps or were only taken in small numbers. These difficulties were overcome in South Africa (Jupp, 1978) by anaesthetizing the bait, in this case baboon or vervet monkey, and roping it to a bait platform, one at 1 m and the other at 10 m above ground. Under the platform two rubber-bladed suction fans in a cylinder sucked into collecting cages any mosquitoes approaching the cage (Figure 8.3). In this instance the exposed and completely unobstructed bait animal attracted and trapped large numbers of the culicine mosquito population in the area, at both levels, compared with zero captures in the unbaited control.

8.3.4   The use of electrified fences for trapping tsetse flies

The use of electrified fences in conjunction with animal models and visually attractive traps has already been introduced in section 7.3.2 with particular reference to tsetse flies. In this section their use and operation

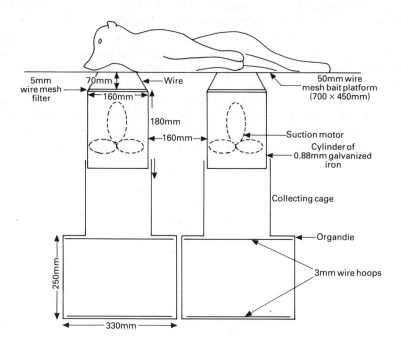

Figure 8.3 *Trap for collecting mosquitoes attracted to anaesthetized monkey or baboon (Jupp, 1978).*

in conjunction with live animal bait will be examined. This technique has been explored and developed to the greatest extent in the continuing studies on tsetse flies in Africa, and it is that work which has provided the bulk of our knowledge.

(a) Development of standard method for oxen and game animals

Having established a standard procedure with regard to the model attractant, and the use and operation of electrocuting surfaces, live animal bait was now introduced into the experimental setup in the form of both domestic stock — mainly oxen — and game animals (Vale, 1974c). For stationary tests, baits were placed in a pen 2.4 m high and 3.6 m in diameter surrounded by a ring of large electrified nets. For mobile bait tests the animals were put in a steel mesh cage or 'crush' mounted on wheels (Figure 8.4). In the experiments described, the investigators concerned have been careful to point out that while many tests make a sharp distinction between mobile and stationary condition of the bait, these really refer to two extremes of what happens in nature, where each individual bait animal will be stationary for part of the time and mobile at others. The experiments with electric netting encircling non-human bait under mobile and stationary conditions indicated that stationary baits attract the most representative cross-section of sexes and species, while mobile baits give the best representation of nutritional states. The flies attracted to mobile baits are generally more replete than those visiting stationary baits.

(b) Experiments to determine the proportion of attracted flies which actually feed on the host

One of the long-standing problems in tsetse behaviour is the question of what proportion of tsetse actually engorge after visiting hosts. This has a

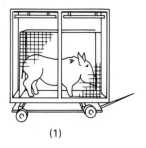
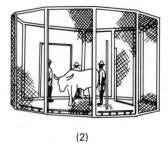

(1)   (2)

Figure 8.4 *(1) Mobile crush and electric cage with warthog in tsetse, studies. (2) Electric pen baited with ox or man (Vale, 1974c).*

bearing on many epidemiological problems, on the diet of tsetse, and on their abundance in relation to the availability of food. The development of the electric fence or electric net has enabled this question to be tackled afresh from a controlled experimental angle (Vale, 1977a,b). A bait animal is placed in a hexagonal pen surrounded by a complete or incomplete ring of electric nets, usually for a 3–4 h period before sunset. The nets measured 3.3 × 1.5 m and could be arranged either lengthways or upright. The arrangement with incomplete netting is shown in Figure 8.5. This was designed to trap and stun a sample of the tsetse flies approaching the bait from outside, and also to deal in a similar manner with samples of tsetse leaving the bait. Flying tsetse do not see these fine nets in time, and when they collide they are killed or stunned and fall into a sticky trap at ground level. In contrast, visual and olfactory stimuli pass through the net unimpeded.

Various possible interfering factors were carefully checked, including the possible visual or obstructive effect of the frame struts or suspensory equipment. In addition, particular attention was paid to the important question of tsetse flight paths in relation to the height of the intercepting nets. This latter point was tested by placing against the upright net (effective height 3.2 m) a series of narrow trays made of transparent fibreglass to intercept and recover flies electrocuted at heights of

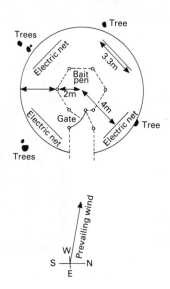

Figure 8.5 *Hexagonal pen with bait animal surrounded by incomplete ring of electrocuting nets (Vale, 1977b).*

20−94 cm, 95−160 cm, 170−244 cm, and 245−320 cm. These were in addition to the wide tray at ground level which recovered all flies below 20 cm. All trays were treated with sticky adhesive.

The results showed that a very high proportion of the fed flies were caught below 95 cm, confirming previous visual observations by many workers that, after feeding, the flight path of tsetse is low and towards the ground or low vegetation. The results also showed that the flight of unfed flies, leaving the bait, maintained much the same height as towards the bait. From the rate at which unfed flies decrease with increasing height, it appears that only a few unfed flies travel at a higher level than 3.2 m, and this is supported by observations made at greater heights elsewhere.

These observations on height of flight path were a necessary preliminary to ensure that really valid samples of approaching and departing tsetse were successfully intercepted, and that any loss, other than through the known and measured gaps between the electric nets, was negligible.

(c) Differential attraction to, and feeding on, range of natural host animals

Single mammals of ten species were exposed in this experimental layout over a series of exposures. The two important figures required were, first, the number of flies that initially approach the bait, and secondly, the proportion of those flies that feed. Considerable variations in attraction were recorded, in which size of host appeared to be an important but not decisive factor. Donkey and kudu for example, provisionally grouped among the larger mammals, recorded a mean of 494 flies as compared with a mean of 487 for the five smaller mammals, sheep, goat, impala, bushpig and bushbuck. The high catches of 750 and 926 recorded on the two largest mammals, ox and buffalo respectively, were exceeded by the catch of 1081 on warthog, an animal only about a quarter their size.

With regard to the proportion successfully feeding, the differences are much more striking, with the highest proportion on donkey, kudu and ox (0.35−0.51), and the lowest proportion (0.00−0.02) on impala, sheep, goat and bushbuck. The finding that there may be a wide gap between the attraction of a particular animal to tsetse, and the extent to which tsetse can feed successfully on that host, is strongly influenced by host behaviour. All the smaller animals with low tsetse feeding rates are very restless and intolerant of insect attack, constrasting with the more placid behaviour of ox and donkey (Vale, 1974a).

This series of experiments also provided a further means of clarifying the already noted depressing effect of man's presence on the attraction of ox to tsetse. Where ox alone was compared with three accompanying men, the presence of men halved the attraction of the ox, and furthermore had the effect of reducing the proportion of flies successfully feeding by about three-quarters.

When the experiments with ox and electric fences was extended to cover all months of the year, they embraced seasons characterized by wide differences in tsetse abundance. In general, the proportion of tsetse successfully feeding was at its lowest at the season of highest fly density, the warm dry months of July to September, and rose to a peak in the late rains and early dry season of March to June. For both sexes of *Glossina pallidipes* as well as for female *G. morsitans*, but not males, there appears to be an inverse relationship between the density of tsetse at bait, and the proportion successfully feeding. This implies that high fly densities may reduce the readiness with which tsetse engorge.

8.3.5 Electrified back pack for human bait in tsetse surveys

The potentialities of the electric net technique in combination with a live host have also been fully explored when man is the host. Originating in the pioneer work in Zimbabwe (Vale, 1974c) this technique has been tested and improved in many other tsetse areas of Africa. Work in Ethiopia for example has been directed towards reducing the weight of the back-pack equipment — including the 12 V battery — while maintaining the voltage necessary to stun tsetse (Rogers and Smith, 1977; Rogers and Randolph, 1978). For use in remote areas it was also necessary to keep the power requirements to a minimum. This was achieved by means of a capacitor which acts as an energy store, and which is discharged only when the fly lands on the grid and is given a powerful shock. Trials on a subspecies of *G. morsitans*, in which the trap was worn on the back of field assistants moving along a set path and stopping at intervals, confirmed that without the capacitor flies which made only a fleeting visit to the grid were able to escape. The new circuit provided a light and efficient alternative to the original design. Although perhaps slightly less effective in stunning flies, the reduction in combined weight of circuit and 12 V battery to 1 kg had great advantages in portability.

When this work was extended to West Africa and the different group of tsetse in that area, namely *Glossina palpalis* and *G. tachinoides*, the use of the electric back-pack on moving human bait was augmented by an electric screen, 60 × 90 cm mounted vertically on four wheels with a towing handle, which was pulled immediately behind the wearer of the pack. The idea is that the back-pack tended to catch flies preparing to land on the human bait above the waist, while the electric screen sampled flies approaching at a lower level (Rogers and Randolph, 1978). In tests, the electric trap on average caught higher numbers of both males and females of these two species in moving than stationary tests, where the back-pack captures were little different from catches obtained by men using hand nets.

In trials in Nigeria, the size of the catching surface was increased, and the colour changed to dark blue (section 7.3.5b) in order to increase the attraction to *Glossina palpalis* (Koch and Spielberger, 1979). In this trial the pack was carried on the back of one man on the fly round, while the other collector followed on behind and watched for any stunned flies trying to escape. These observations indicated that for this species at least, the electric back-pack was most efficient at high fly densities, but relatively inefficient at very low tsetse densities compared to other sampling methods. Some idea of fly densities encountered in these trials can be gained from the fact that in 'high density' areas, ten flies were recorded per 1 km, while in 'low density' areas, this figure fell to 0.3.

Among the advantages of this technique compared with others such as the biconical trap and hand catching was the finding that electric back-packs caught a higher proportion of recently fed females. This is important in view of the fact that recently engorged females are not easy to detect in their natural habitat at these densities, making it difficult to obtain adequate samples for blood meal testing. Among the disadvantages was the liability of damage to equipment in rough field conditions. It was also observed that many of the flies appeared to be only stunned by the electric grid, and managed to escape from the collecting box.

8.4 Trapping based on odour responses of tsetse flies

8.4.1 Interrelation between visual and olfactory responses

One of the most significant facts emerging from the intensive experimental work on tsetse in the 1970s was the fact that while the visual element of traps played a key role in attracting tsetse flies to moving targets, and in the final guidance of tsetse to static baits or targets, overall, bait odour was the major factor attracting tsetse to live host animals. For example, in the course of an afternoon the odour from a single ox concealed in a covered pit could attract up to 1500 *Glossina morsitans* and *G. pallidipes*, while catches in no-odour traps, i.e. with visual element alone, were found to be only 7−11% of that number (Vale, 1977b).

At this stage the trend of research was increasingly orientated to the practical possibility of controlling, as well as monitoring, tsetse populations by means of odour attractants which would bring about a concentration of flies near traps treated either with insecticide or with sterilizing devices

(Vale *et al.*, 1988). From that time onwards, continuing through the 1980s, there has been a concerted effort on the part of scientists, both those involved in field work in Africa and those dealing with laboratory assay work in the UK, to analyse and identify the components of these all-important odours (Vale, 1979).

The problem of the nature of the attractive components of host odours in general is long standing, affecting all blood-sucking flies, and has been the subject of particular attention in mosquito research over the years. One of the more obvious components of these host odours, whether from the body or the breath, is CO_2, whose importance is well established, and which is now widely used as an added attractant in a whole range of trapping devices covering a wide variety of blood sucking insects (Gillies, 1980). Until the intensive work on tsetse had gained momentum, however, very little progress had been made beyond that stage in other fields. The outcome of this remarkable progress in tsetse studies is therefore by no means of relevance only to that group, but must inevitably influence all research workers concerned with this fundamental problem elsewhere.

8.4.2 Design of traps in odour experiments

In field testing of odour components in Zimbabwe (Vale, 1980) the experimental trap most widely used has three components. The first is a visual target to assist attracted tsetse to locate the odour source precisely; this visual element took the form of a black cylinder, 50 cm long, 37 cm in diameter, mounted horizontally 37 cm above the ground, and sited 50 cm downwind of the odour source. The second component was an electric net 1×1 m placed 1 m downwind of the odour source, and the third component was the odour source or dispenser. In addition to the various odour components to be tested, the trials involved natural ox odour obtained by placing a single ox in a roofed pit and drawing the odour to ground level through a ventilated shaft fitted with an electric fan. In some experiments the natural odour was first passed through a charcoal filter.

8.4.3 Tests with odour components

Among the first likely components of ox odour to exhibit attractant properties were CO_2, acetone, and certain aldehydes and ketones (Vale, 1979, 1980). This was followed by the discovery that in the laboratory one of the recently identified odour components, octenol (1-octen-3-ol), is a powerful olfactory stimulant capable of inducing upwind flight of tsetse (Hall *et al.*, 1984; Bursell, 1984). From then onwards, field tests concentrated on the role of this compound, either on its own or in various blends with other established chemicals such as CO_2 and acetone. In addition, levels of acetone and octenol were monitored in the air drawn from pits containing an ox (Vale and Hall, 1985). Compared with other more volatile chemical components of odour, octenol is one of the 'not-so-volatile' components, and normally emanates from the ox at rates below $0.05\,mg\,h^{-1}$.

When the ox odour was passed through a charcoal filter to remove these not-so-volatiles, including octenol, catches of *Glossina pallidipes* fell by 50%, but not so markedly with *G. morsitans*. It appears that octenol could account largely or entirely for the efficacy of this fraction of not-so-volatiles, but that in the case of *G. pallidipes* there must be another attractant ingredient in addition to octenol.

When catches obtained with ox odour alone were compared with those obtained with combinations of ox odour and a range of concentrations of octenol, addition of octenol at the rate of $0.05\,mg\,h^{-1}$ had no effect, presumably because it merely doubled the normal rate of emanation. However, the addition of concentrations of $0.5\,mg\,h^{-1}$ increased catches by a factor of about 1.5 for *G. pallidipes* and 2.5 for *G. morsitans*. Further increases in the octenol level were found to decrease the catch, especially for *G. pallidipes*, and when the level was increased to 500 mg it became strongly repellent. Further studies were carried out to determine if attraction to ox odour could be explained simply by a response to CO_2 acetone and octenol. Various combinations of these three were tested, and it was established that these artificial mixtures are highly attractive on their own, and that some of these mixtures are significantly more attractive than natural one-ox odour. However, it became evident that other important components, still unidentified, may exist, possibly among the ketones, such as butanone, aldehydes or fatty acids, and that some of these unspecified chemicals might possibly act by synergizing the action of octenol.

8.4.4 Tests on undetermined ox odour ingredients

On the question of undetermined ox odour ingredients, research has also been carried out in another direction, in particular the host residues associated with ox urine and bush pig bedding (Vale *et al.*, 1986). The trap design differed in detail from that described above in that it consisted of a visual bait of black cloth placed beside two sheets of fine black netting. This utilizes the principle that many of the tsetse attracted to the black cloth, but flying around it, will collide with the fine black net. Experimentally, both sides of cloth and net are covered by a grid of fine electrocuting wires, but in practical field control these surfaces would be impregnated with insecticide. All odour sources were placed on the ground 30 cm downwind of the trap.

The main attractants tested were buffalo and ox urine, and sacks on which bush pig had slept. Catches of *G. pallidipes* were found to be increased by 73% with the ox urine, and 77% with buffalo urine, in contrast to the pig sacks which exerted no odour attraction. The effect of urine seems to be similar to that of CO_2, acetone, butanol and octenol, and is quite distinct from the full mixture of natural host odours. Increase in the concentration of natural host odours from that produced by one ox to the output of six oxen in a covered pit had already been found to increase the catch of *G. morsitans* and *G. pallidipes* from 1500 in the course of an afternoon to 4000 (Vale, 1977b). A further massive concentration of natural host odour by progressively increasing bait masses from 500 kg, to 3500, 6500 and 9500 kg, culminating in a maximum mass of 11 500 kg produced by a combination of 37 cattle, 22 goats, 43 sheep, a donkey and a buffalo, showed that catches of all tsetse, except male *G. morsitans*, increased considerably with increases in bait mass in excess of 3500 kg (Hargrove and Vale, 1978). There is no indication that the concentration of natural host odours ever reaches a point where it becomes repellent. On this issue therefore, the effect is in marked contrast to that of many of the individual odour components where increasing concentration eventually reaches a point where attraction gives way to repellence.

The question of identification and isolation of the various chemical components of cattle urine has been accompanied by studies on their effect on tsetse behaviour in laboratory wind tunnels (Bursell, 1987; Bursell *et al.*, 1988). Field testing was carried out by placing the odour component on the ground 30 cm downwind of the conventional three-dimensional tsetse trap — Beta and F-3 designs. Some of the synthetic components were also tested with the electrocuting target described above.

The attraction of cattle urine was shown to be entirely attributable to the phenol compounds which it contains. Four of the eight naturally occurring phenol derivatives produced tsetse response in the laboratory and induced upwind flight in wind tunnel bioassays, and were also found to increase trap catches in the field.

In this integrated research programme there was good accord between tsetse response observed under laboratory conditions, and the behaviour of flies in the field, even though the conditions of testing differed on some important points. In the wind tunnel, odours were routinely administered in 0.2% CO_2 which had an activating effect, thus ensuring reasonable frequency of take off. In the field tests, odours were all tested in the presence of acetone and octenol. These different test environments may account for one apparent discrepancy in that 2-methoxyphenol, one of the minor members of the eight naturally occurring phenol derivatives, induced upwind flight under laboratory conditions, but caused a reduction in trap catch in the field. This may be accounted for by the fact that the laboratory assay concentrated on single elements of behaviour, while in the field a multitude of interacting responses — not all fully understood — are involved. As none of the phenols tested in this series proved to be as attractant as urine itself, it seems possible that the effects of single phenols may be determined by interaction with other phenols.

8.5 Wind tunnel experiments on effect of CO_2 and other odour components on *Stomoxys*

The important part played by wind tunnels and other laboratory observation equipment in analysing odour responses of blood-sucking flies is well illustrated by work on an allied biting fly, *Stomoxys calcitrans*, the stable fly (Warnes and Finlayson, 1985a,b). The activity of *Stomoxys* was investigated by recording the number of flights made by individual flies in a flight chamber consisting of a nylon mesh cage within a second box of metal and perspex, provided with an observation port. In this, filtered air passed from left to right at the rate of $841 \, \mathrm{min}^{-1}$. CO_2 could be added to the airstream from a cylinder and needle valve. The flight chamber was modified by attaching a target in the form of a disc of black filter paper, 7.5 cm in diameter, to one wall. Records were made of the number of landings on the target and on the rest of the chamber over an experimental period during which the CO_2 concentration was increased.

The initial response to increased CO_2 was an increase in undirected flight activity, which led to an increase in upwind flight. With the addition of expired breath, an increase in flies in the upwind section was produced, greater than the equivalent increase in CO_2. In previous studies on the same issue it had been concluded that the effect of expired breath on *Stomoxys* could be accounted for on the basis of CO_2 alone. This more recent work shows that the odours present in human breath, and CO_2, are synergistic. The work on tsetse flies had shown that another component of human breath, acetone vapour, also appears to act in a synergistic manner, increasing the catch of *Glossina* in the field (Vale, 1980). A similar effect was produced by acetone in the laboratory tests with *Stomoxys* where it acted by eliciting upwind flight.

In the field trials on tsetse flies in Africa, many observations were also made on *Stomoxys* taken in the traps, and these had shown that in contrast to acetone, addition of acetic acid to the traps reduced catches of this fly. This was substantiated by the laboratory tests which showed that the addition of acetic acid to the air stream in the observation chamber reduces the fly's response to increase in CO_2 concentration.

With regard to the visual response of *Stomoxys* in the observation chamber, the increased activity resulting from the increased CO_2 concentration leads to flight; this in turn leads to increased numbers of landings on the target. However, the increase in CO_2 does not elicit visually directed flight, the increased landing rate on the visual target being simply a reflection of the increased landing rate consequent on the increased take-off.

8.6 Odour responses of blackflies (*Simulium*)

Of the many observations carried out on host response of blackflies in many countries, one is of particular significance because the experimental approach adopted had developed, quite independently, along lines which are similar on many essential points to those used in the tsetse research programme described. In this case the bait animal was man himself, the investigation being prompted by the fact that in West Africa strong attraction to man on the part of key members of the *Simulium damnosum* complex not only determines the role of these biting flies in the transmission of human onchocerciasis or 'river blindness', but also dictates that the attack rate on man as judged by 'flies per boy/hour' is still the main method for monitoring fly populations, and following their seasonal fluctuations in density and composition. It was hoped that these studies, carried out in the Cameroons, might pinpoint certain vital factors which could be put to practical use for the design of a mechanical trap to replace the complete dependence on live human bait as the attractant (Thompson, 1976a,b).

Experiments were designed to evaluate the relative importance of visual stimuli (sight), olfactory stimuli contained in human breath (exhaled breath), and olfactory stimuli emanating from the skin surface (smell). The problem was approached in two ways: first, by presenting as bait a man emitting all the attractant factors except the one under consideration; and secondly, by attempting to isolate as far as possible, and present as bait, the factor under consideration. In this method of experimental approach this investigation shares many of the features of the concurrent studies on tsetse in Rhodesia (Zimbabwe) at that time, but it is clear that the methods were arrived at quite independently.

Traps were specially designed for this purpose. The first of these was a triangular frame 120 cm high, and 44 cm each side (Figure 8.6A). Two

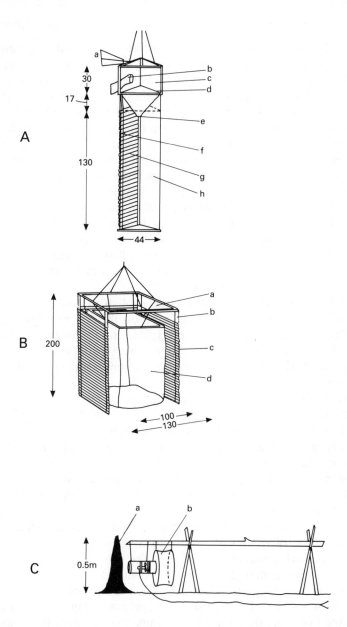

Figure 8.6 *Traps designed to study reactions of* Simulium *to visual and olfactory components of baited traps. A. Slat trap; (a, rudder; b, sleeve; c, fine mesh ('sandfly') netting; d, 10-mm entry gap; e, weight; f, slat holder; g, slats (43 × 5 cm), h, 1-mm mesh netting.) B. Enclosure trap, (a, 1-mm mesh netting; b, polythene (transparent); c, slats (130 × 5 cm).) and C. Fan trap (a, bait cloth; b, collecting cage) (Thompson, 1976a).*

sides of the column were covered with fine mesh, while the third was provided with a close series of inclined slats. The trap was suspended from the branch of a tree with the open bottom 10 cm above the ground. The operation of the trap was based on the belief that flies would travel upwind towards the bait inside the trap, and finally land on the slats. Moving upwards they would reach the top of the slats and then find themselves inside the column where they would be attracted to the two net-covered sides through which the daylight penetrated. From there they would continue to move upwards towards the top of the column. This column was topped by a netting cage constructed in such a way as to leave a 10 mm gap between the outside wall and an inverted net tetrahedron suspended in position with twine. This downward pointing tetrahedron was kept in a taut position by means of a small weight, ensuring that the narrow 10 mm gap was kept rigidly open to allow ascending flies to move upwards into the cage, where they were trapped and could be removed through side sleeves.

The second trap (Figure 8.6B), large enough to contain a man, was called the enclosure trap, and was based on the slat principle with the slats adjusted so that the man was not visible from outside. The trap was used to observe the number of flies approaching a man emitting smell and exhaled breath stimuli, either separately or combined, in the absence of visual stimuli.

In the first series of experiments with the slat trap, comparisons were made between (i) the empty trap, (ii) a trap provided, about 20 cms in front of the slats, with a rubber hose outlet through which a man seated 30 m away in thick bush could exhale, (iii) a trap provided in front of the slats with recently worn trousers, and (iv) a trap provided with both exhaling hose and trousers. Over the same period two collectors 100 m downstream made conventional man-bait collections of blackfly. The results are shown in Table 8.1.

From these figures it can be seen that exhaled breath and worn trousers on their own attracted no more flies than the empty trap, but that the catch was increased roughly seven-fold when exhaled breath and trousers were presented together. Even with this combination, the catch was less than 10% of the comparison man-bait catch in the open.

In a second series of tests, using the enclosure trap, a human subject not visible from outside sat inside and compared his attraction under two different conditions: first, exhaling normally, and secondly exhaling through a rubber tube with its outlet 30 m away in thick bush. The results showed that when visual stimuli are reduced, human smell alone attracted four times as many flies as the unbaited trap, but when exhaled breath and smell were combined, the catch increased by a further factor of four.

Table 8.1 Numbers of *Simulium damnosum* captured over 12 hours in slat traps with various baits on successive days

	No bait	Exhaled breath	Worn trousers	Exhaled breath and worn trousers
Trap catch	54	42	30	256
Vector collector catch	3365	2728	3224	2652
Trap catch as proportion (%) of vector collector catch	1.6	1.5	0.9	9.7

(Thompson, 1976).

In a third series, the attraction of human exhaled breath was tested when visibility of the human bait was optimal, in contrast to the test above. This was done by one of the exposed bait men exhaling through a rubber tube with its outlet 50 m away. The results showed that the removal of exhaled breath did not remove the attraction of fully-visible humans. These results were found to apply in both the forest and savanna environments in which these tests were carried out.

A further experiment, using a fan trap, was designed to study the attraction of a human bait when fully visible and moving round from time to time, as compared with one stationary and partly concealed (Figure 8.6C). In each case a further comparison was made between bait exhaling normally and bait exhaling through a rubber tube or hose to a distant exit. These experiments showed that the hidden stationary man always attracted fewer flies than the moving exposed man, irrespective of whether or not they were exhaling normally.

In further experiments, using a transparent plastic enclosure, the human host was easily visible, but the exhaled breath and body odours were confined to the airtight enclosure. At a time when biting density outside was of the order of 1000–2000 flies per hour, the visible but odourless bait in the plastic enclosure was completely non-attractive.

By means of the fan trap, various choices of washed versus unwashed garments were presented as stimuli, and these showed that worn clothes attracted over ten times as many flies as unworn clothes. In both cases the attraction was greatly enhanced by the addition of CO_2, up to 18 times greater in the case of the worn clothes. The findings on the strong attraction of human body odours also provided an explanation of the observation that fan traps which had been used for several consecutive days showed higher and higher catches in the controls. By wearing gloves and protective clothing, and washing and sterilizing the equipment, it was shown conclusively that high control catches stemmed from handling contamination.

These experiments also revealed differences between the reactions of flies in the Sudan savanna zone of the Cameroon, and the forest zone. In the savanna, the removal of exhaled breath and the masking of human odour failed to produce any substantial reduction in the number of flies attracted to human subjects. This suggests that in that zone, some other stimulus, probably sight, is a major factor in attraction. In this case it seems very likely that the differences in behaviour reaction in the two zones could be attributed to the likelihood that two different members of the *Simulium damnosum* complex were involved.

These and other results suggest that different members of this complex may, in their host-locating activities, respond to different kinds of stimuli emitted by the human host. The major role of human body odours as attractants in the forest zone experiments was further analysed by comparing the attractiveness of clothing worn on different parts of the body, using the fan trap technique. In these tests trousers were found to be 5.5 times more attractive than shirts. This led to an examination of the attraction of various body fluids and exudates, of which sweat emerged as the major attractant, particularly sweat from below the human waist line. Attempts to isolate and identify these particularly attractive chemical compounds in human sweat failed to pinpoint the active ingredient, possibly because such chemicals did not react with the range of organic solvents used in extraction. Whatever this compound is, flies seem to be able to detect extremely small quantities as evinced by the experiments with fan traps 'contaminated' by human handling. Should these studies lead eventually to the extraction of this powerful compound, or compounds, then there would be a real possibility of using that substance to design a standard trap which would obviate the continuing need for human bait and man-biting indices as the sole measure of *Simulium* activity and abundance.

Chapter 9

Attraction of Blowflies and their Allies to Carrion-Based Traps

9.1   Introduction: The Australian sheep blowfly (*Lucilia cuprina*) and the American screwworm fly (*Cochliomyia* spp)

Among the large group of non-blood sucking blowflies, most of whose members breed in carrion and are not economically important, there are several different species in different parts of the world whose habits render them important pests, mainly because of the injury they can inflict on domestic animals. One of these, the sheep maggot fly (*Lucilia sericata*), is a well known pest in Britain. There are two other regions of the world where the habit of depositing eggs or larvae on live animals poses an even greater threat, necessitating continuous research into the habits and control of these insects. The Australian sheep blowfly (*Lucilia cuprina*) is mainly responsible for sheep myiasis in that continent. In the United States the main allied problem is caused by the screwworm flies *Cochliomyia hominivorax* and *C. macellaria*, primary causes of myiasis in cattle.

In both regions there has been intensive research over the years, one of whose objectives has been the development of traps for monitoring and controlling those blowflies. The basis of much of this research is the fact that, despite their specific association with live hosts, these species are attracted to meat or carrion baits, and to the products of decomposing meat. Starting with that common factor, later courses of study in these two regions have developed quite independently, leading to distinctly different approaches to a common problem. The methods developed in these two areas of trap design, nature of bait and trap evaluation are very illuminating, particularly as a major common bond arose in designing traps which could be used to implement genetic control based on auto-radiation of trapped flies (Vogt *et al.*, 1985b); Wardhaugh *et al.*, 1983).

The study and trapping methods developed for the Australian sheep blowfly have proved of great value in the study of another Australian non-blood sucking pest fly, the bushfly *Musca vetustissima*. The bushfly is not a blowfly and it breeds in cow dung. Its main economic importance is as a nuisance fly which tends to swarm around and settle on the human face, constituting a serious problem in many parts of the country. Experience gained in trapping blowflies and in analysing trap reactions has been responsible for the marked progress in trapping and monitoring bushfly populations as well.

9.2 The Australian sheep blowfly

9.2.1 History of carrion-baited traps

The Australian sheep blowfly rarely breeds in carrion due to intense competition from native calliphorines; nevertheless, it is attracted to dead meat and carrion, and consequently carrion-based traps have long provided the obvious, simplest and most efficient means of sampling fly populations, and estimating seasonal changes in fly abundance. Experience of such prototypes widely used in the 1940s, highlighted certain shortcomings in design and performance, and this led to the development of an improved design of 'West Australian blowfly trap' which has now become the accepted standard since the mid-1970s (Vogt and Havenstein, 1974). This new trap (Figure 9.1) deals with certain shortcomings of earlier designs by (i) excluding flies from the bait pan by means of a screen; (ii) incorporating

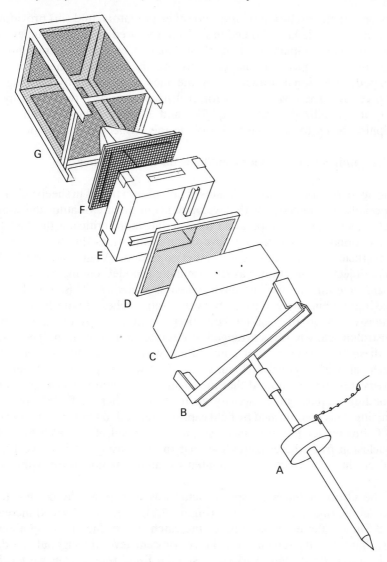

Figure 9.1 *Improved design of 'West Australian blowfly trap'. A. Oil container (for ant exclusion). B. Trap stand. C. Bait pan. D. Excluder (to prevent flies reaching bait). E. Entry chamber. F. Cone, pyramid of copper gauze. G. Fly chamber (34 × 25 cm) (Vogt and Havenstein, 1974).*

devices for the exclusion of ants and other predators, and (iii) providing a screen to exclude large-bodied insects not required in blowfly studies. The basic principle remains unchanged in that flies entering the trap through the slits, then pass via the cone to the fly chamber where they are trapped. Additional attraction for the flies is provided by painting the trap yellow. The essential bait for each trap comprises two sheep livers, 20 g of crystallized sodium sulphide, and 1 l water, the function of the sulphide being to enhance and prolong the attraction of the bait.

9.2.2 Influence of weather conditions on trap efficiency

The need for more precise estimates of population trends in sheep blowflies, especially in connection with developments in the use of autocidal control techniques, made it imperative to examine more critically the fact that conventional trap catch data have certain limitations due to variables in trap efficiency under different weather conditions (Vogt et al., 1983). The main object of this study was to formulate a model relating catch rates of Lucilia cuprina to various environmental variables. This was done by studying within-day variations between trap catches of wild flies over an extensive range of weather conditions, using groups of 12–18 West Australian blowfly traps arranged at intervals of 0.5–1.0 km. The results of these trials showed that temperature was the main variable affecting catch rates of L. cuprina (Figure 9.2). This was supported by laboratory observation which showed that increasing temperature produced a corresponding increase in fly activity up to 26.7°C where it flattens out to a plateau. This is confirmed by field captures related to temperature; beyond 38°C further temperature increase produces a fall off in catch rate. The conclusion from these figures was that the increasing catch above 15.5°C reflects increased activity stimulated by higher temperature within that range.

The effect of the wind speed factor was to depress the catch rate of Lucilia cuprina above 2.5 m s^{-1} (Figure 9.3). When wind speed increases to 8.5 m s^{-1} the effect is to halve the catch. Other factors being allowed for, flight activity remains at a fairly constant level throughout the day.

To allow for all these variables, 'standardized' trap catches were calculated to represent relative measures of population size which differ from actual population size by a constant scaling factor. In arriving at these catch rate models it was assumed that both sexes reacted similarly to environmental variables. The sex ratio on emergence is 1:1, but in trap catches females invariably outnumber males. As this bias could represent differences in behaviour, or reflect differences in mortality, experiments were designed in which baited traps were cleared four times a day to

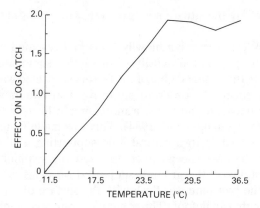

Figure 9.2 *Effect of temperature on log catch rate of* Lucilia cuprina, *after removal of effects due to other environmental variables (Vogt* et al., *1983).*

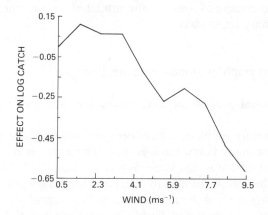

Figure 9.3 *Effect of wind on log catch rate of* L. cuprina *after removal of effects due to other environmental variables (Vogt* et al., *1983).*

cover four periods, dawn−9, 9−12, 12−15, and 15 hours to dusk (Vogt *et al.*, 1985c). Differences in response of the two sexes to four environmental variables were also tested, temperature, wind speed, relative humidity and solar radiation. The radiation effect was found to be highly significant and reasonably consistent, males responding more strongly than females at high radiation levels.

The time-of-day records also revealed differences in sexual response. During the first trapping period of the day, male response is very low, but later in the day the situation is reversed, with males more active than females.

The use of West Australian blowfly traps had revealed marked spatial variations in the population density of flies, with some sites consistently more favourable than others. One of the factors suspected was the presence or absence of sheep. In order to study fly distribution within smaller areas, such as between paddocks, a more simple liver-baited sticky trap was designed (Wardhaugh *et al*, 1984). This was made from 15 cm square white painted board, with a central hole supporting a 20 ml plastic vial containing the bait — sheeps liver and sodium sulphide. Traps were placed horizontally on the ground, supported on twigs or stone, so that the mouth of the bait tube was flush with the surface of the board, which was smeared with Tanglefoot. Three parallel transects, each of ten traps, were laid out in each paddock, with traps and transects at 30 m intervals covering an area of about 5.6 ha. Most flies were caught in the sheep compounds as compared with the adjacent cattle campsites. These traps, simple and inexpensive to operate, also attracted several other calliphorine flies of veterinary importance.

9.3 Trapping bushfly *Musca vetustissima*

9.3.1 Traditional methods of hand-netting flies attracted to man

Bushfly and bushfly populations have been the subject of intensive research in Australia for many years, but it is only comparatively recently that the great potential of traps has been fully exploited. For many years the attraction of this pest fly to human sweat and secretions provided the only feasible catching method, by netting flies as they approached human bait. This was the basis of studies on bushfly populations initiated over 20 years ago, mainly concerned with the problem of bushfly disappearance from southern areas of the continent in autumn and their reappearance in spring (Hughes, 1970, 1974). The limitations of hand-netting as a basis for population estimates was early recognized, but attempts to devise an alternative method using baited traps was discouraging. Western Australian types of trap — in their prototype form — baited with ground sheep's liver and sodium sulphide, were found to have only a low attraction to bushfly (Norris, 1966), and this was also the experience with the improved design of Western Australian trap (Vogt and Havenstein, 1974). However, the potentialities of these traps for bushflies had to be drastically revised and re-examined in the light of new information about the role of bait age in attracting flies.

9.3.2 Influence of bait age

It was already recognized that in the sheep blowfly, the baits became more attractive with increasing time of exposure, with marked differences in catch between fresh and 7 day old, well decomposed liver-based bait. However, when monitoring population trends, the practice adopted was to discard baits after 3 days in case the increase in attraction could be wrongly interpreted as indicative of increase in fly population (Vogt *et al.*, 1983). In the course of routine trapping of *Lucilia*, it was observed that bait aged a week or more became very attractive to *Musca vetustissima*, which were virtually absent from fresh meat. This crucial bait-age factor dictated a completely new approach to the role of trapping in bushfly studies. The preference for older baits as compared with new was found to vary from 4.5:1 to over 24:1, this variability probably being a reflection of changing levels of fly activity (Vogt *et al.*, 1981, 1983). Females are more trappable than males, forming 80% of the catch, and of these females newly emerged are much less attracted than those in mid-egg stage.

9.3.3 Development of wind-orientated trap

A further step in progress was the development of a wind-orientated trap designed to overcome some of the disadvantages of the standard usage of the static Western Australian design (Vogt *et al.*, 1985a). This trap (Figure 9.4) consists of a separate bait chamber and fly chamber. Traps were operated daily for six hours, and flies collected every hour. Further

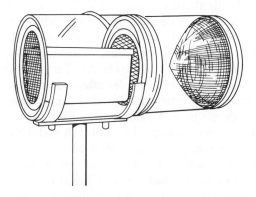

Figure 9.4 *Wind-orientated trap for sampling bushfly populations,* Musca vetustissima, *mounted on vane 1.5 m above ground (Vogt* et al., *1985a).*

investigation was made into the vital factor of bait composition and bait age.

In this series the bait consisted of minced sheep's liver and sodium sulphide solution. With the addition of fresh cattle dung, which by itself is well known to be attractive to bushflies both as an ovipositing site and a source of protein, various combinations of cattle dung, liver and sodium sulphide were compared experimentally. In addition, blowfly baits were aged by introducing several hundred blowfly larvae. Cattle dung and blowfly baits did not differ significantly in their attraction to bushfly, but were considerably more attractive when presented together, either mixed or side by side, catching anything up to 600–700 flies per trap per day.

Crust formation on bait accounted for variations in bait attraction, and this was dealt with either by hourly stirring or by adding blowfly larvae whose activities prevented crust formation. A potential variable in trap effectiveness was trap saturation caused by the presence of large numbers of flies in the fly chamber, up to 1000 bushfly per hour on occasions.

9.3.4 Influence of temperature factor, solar radiation and trap height

Comparison field collections of bushfly made by the traditional method of hand-netting flies as they approach the human bait/collector show that *M. vetustissima* is more highly attracted to the bait in the wind-orientated trap than to people, with trap catches 3.6 times higher on average than corresponding net catches. The rate at which *Musca vetustissima* is netted has long been known to be influenced by temperature, and the effects of that factor were taken into account in determining the 'abundance index' of flies caught per man on the human bait/catcher, which is assumed to be proportional to the fly density.

This temperature factor — the main weather variable — was re-examined in the baited wind-orientated trap (Vogt, 1986), based on observations on 35 trapping days on which the traps were cleared at two hourly intervals throughout the day from 8.00 to 18.00 hours. Temperature was found to account for 71% of the within-day deviations of the two hourly catch, bearing in mind that flight activity of *M. vetustissima* ceases below 12.5°C. The optimum response to traps was found to occur at 27.5°C. Relative humidity and wind speed accounted for only 0.8 and 0.9% respectively of the within-day deviance, but solar radiation was revealed as a more important factor, accounting for 11.5% of within-day deviation. Both declining levels of solar radiation, such as towards sunset, and increased cloud cover reduced both trap catches and net catches.

The effect of trap height had also to be considered in relation to determining the best standard for comparative purposes. Wind-orientated traps set up at different heights showed that *M. vetustissima* was caught at all heights up to the maximum of 6 m tested, but that overall the numbers taken were inversely related to height (Vogt, 1988). A breakdown of catch at heights of 1, 2, 3, 4, 5 and 6 m, showed that 85% of the males caught and 91% of the females were taken at levels of 3 m and under. Traps at 1 m caught 41% of the total males and 63% of the total females, providing a useful pointer to establishing a standard height for trap setting.

9.4 Screwworm flies

9.4.1 Carrion bait

The screwworm flies, *Cochliomyia hominivorax* and *C. macellaria* are primary causes of myiasis in cattle in the USA. The eggs deposited on the live host produce larvae which burrow into the skin and cause great damage to the tissues and eventually considerable commercial loss in the hides. Like the allied calliphorines the blowflies, these flies are attracted to carrion bait, and this has formed the basis of trapping and reducing the fly population over the years. The original trap design, the Bishopp trap (Bishopp, 1916), is of simple basic design and is composed of a vertical cylinder of plastic screen with a vertical funnel attached to the bottom rim. Flies are attracted to the bait in a container under the trap, and move up into the cylinder towards the light. Since their first introduction in 1916 these traps, baited with decomposing liver or meat products, were used for over 40 years.

As a method of sampling fly populations, certain limitations of the Bishopp trap have long been recognized. Chief among these is their relatively low attraction to male screwworm flies, the proportion of females to males on liver bait being over 50:1 (Jones *et al.*, 1976). In addition, many of the flies attracted are not trapped as they crawl over the bait but do not move up into the cylinder. Another shortcoming of this trap is that it attracts a large number of non-target or 'trash' flies, making sorting and counting more difficult.

From the early 1970s, considerable effort has been made not only to design a more efficient type of trap for standard use, but also to find an alternative bait to liver, one that would be more attractive to males. The importance of increasing the male capture is emphasized by the increasing use of sterile-fly release strategies in screwworm eradication programmes in the south west USA (Coppedge *et al.*, 1978), and the need for this trapping programme to meet the important objective of early detection of native screwworm.

9.4.2   Development of new wind-orientated design of trap

Observations on fly behaviour towards baits showed that they had a strong flying orientation towards the attractant, and tended to congregate on the trap screen immediately downwind of the bait. This suggested that an improved design would allow flies to enter through a horizontal funnel on the downwind side of the trap, rather than vertically as in the standard Bishopp trap, and that the chemical attractant would be placed upwind of the funnel (Broce *et al.*, 1977). Implicit in this design is that the trap is wind-orientated to ensure that the attractant is maintained in an upwind position. This was achieved by means of two metal vanes (Figure 9.5).

Field trials with the wind-orientated trap and the Bishopp trap showed that both were liable to exhibit considerable day-to-day variations, and that there were significant differences in trap response in the two screwworm

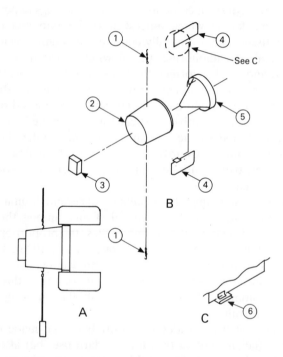

Figure 9.5 *Wind-orientated trap designed for screwworm flies,* Cochliomyia *spp. A. assembled trap with stabilizing weight; B. exploded view of trap; C. detail of wind vane clip (1. Eyehooks and swivels; 2. Plastic pail with screened end; 3. Metal and screened box for SL-2; 4. Wind vane; 5. Screen funnel; 6. Wind vane clip.) (Broce* et al., *1977).*

species involved. In the case of *C. hominivorax*, the wind-orientated trap caught a greater number of both wild and sterile (released) males and of females than the standard trap whereas with *C. macellaria* — which has made up to 65% of the total flies in the standard trap — these now only constituted 29% of the catch in the wind-orientated trap. Many of the *C. macellaria* observed approaching the funnel, hovered at the entrance and then darted out again. In contrast, *C. hominivorax* approaching the funnel, usually went directly to the opening.

Sex ratios of *C. hominivorax* also differed in the two trap designs, the standard design trapping a higher percentage of males (28%) than the wind-orientated trap (17%). This difference was offset by the greater number of males actually trapped in the latter.

9.4.3 Programme of bait improvement: development of chemical baits

Alongside the advances in trap design there have been even more significant developments in bait improvement dictated by the long-felt need to find an alternative to the standard liver bait, and one which was more attractive to males (Jones *et al.*, 1976). This led to a more critical examination of the different chemical components and meat decomposition products, of which about 30 — alcohols, acids and esters — were tested in the standard Bishopp trap. As a result of this, an attractant bait involving ten chemical degradation products was devised which proved as effective as liver, with the added advantage of attracting more males. The proportion of females to males in the liver-baited trap was 53:1, compared with 5.6:1 in the chemical-baited trap.

This combination of ten chemical decomposition products was given the name Swormlure (Coppedge *et al.*, 1977, 1978), the main components being butyl alcohol, several organic acids, phenol, cresol, indole, etc. This original blend of Swormlure, or SL-1 led to the development of a further improvement, SL-2, which contained an additive dimethyl disulphide (Mackley and Brown, 1984). Later, acetone was deleted from SL-2 and this formulation was adopted as the basis for screwworm survey and suppression in the southern USA. When the eradication programme moved into Mexico, erratic performance of the SL-2 dictated the need for further improvement. By experimenting with different proportions of the same basic ingredients, as well as 15 new candidate chemicals, an improved version, Swormlure-4 was developed. This compound differs from SL-2 in that it contains high amounts of the compounds with the four lowest boiling points, and lower amounts of the compounds with the six highest boiling points. It has now replaced SL-2 as the standard attractant for screwworm flies in the eradication programme.

By now, wind-orientated traps had replaced the original standard models in which the earlier tests on chemical bait had been carried out. The continual release of sterile flies provided abundant material for these experiments, supplemented in some areas by wild flies. The function of these traps is not just routinely to monitor populations, but also to determine if sterile flies released from an aeroplane are alive when they reach the ground. The trapping grid was also vital in showing that the trapping pattern and dispersal of ground-released sterile flies was similar to that of wild flies.

9.4.4 Importance of trap siting

Trap siting was found to play an important part in this control and surveillance system. Traps placed in open sunlight or very light shade, protected from direct wind, were known to be more attractive to screwworm than those in shade. This factor was re-examined critically in the warm humid climatic conditions of south Mexico (Welch, 1988). A comparison was made of three different trapping situations: (i) traps suspended from tripods in open pasture; (ii) traps suspended from trees in the same pasture; and (iii) traps suspended from trees within the woods along the edge of the pasture. The adult *Cochliomyia hominivorax* for these trials were provided from cartons of sterile pupae, marked with fluorescent powder, placed out in the field at weekly intervals throughout the experimental period. Nine wind-orientated traps were placed in a circle of diameter 300 m with the release point at the centre, three of these in the woods and six in the pasture. The traps were set at 0.7 m above ground, and were baited with 40 mg of Swormlure-4. An estimated 50 304 flies were released and, of these, 4934 (9.8%) were recaptured. Traps in the open pasture collected more than in the woods. In the pasture itself, traps suspended from isolated trees collected more flies than traps suspended from tripods, possibly due to the fact that traps in isolated trees were in a more shaded position than tripod traps in full sunlight. An additional factor is that isolated trees are attractive aggregation sites for resting flies. The break down of the catch was as follows:

	Males	Females	Males + females
Woods	66	73	139
Trees	1262	1782	2044
Tripods	756	995	1751
Total	2084	2850	3934

References

Adkins, T.R., Ezell, W.B., Sheppard, D.C. and Askey, M.M. (1972). A modified canopy trap for collecting Tabanidae (Diptera), *J. Med. Ent.* **9**, 183–185.

Agee, H.R. and Patterson, R.S. (1983). Spectral sensitivity of stable, face and horn flies, and behavioural responses of stable flies to visual traps (Diptera: Muscidae), *Env. Ent.* **12**, 1823–1828.

Ahmad, T.R. (1988). Field studies on sex pheromone trapping of cotton leafworm *Spodoptera littoralis* (Boisd.) (Lepidoptera: Noctuidae), *J. Appl. Ent.* **105**, 212–215.

Allan, S.A. and Stoffolano, J.G. (1986). Effect of background contrast on visual attraction and orientation of *Tabanus nigrovittatus* (Diptera: Tabanidae), *Env. Ent.* **15**, 689–694.

Allen, D.C., Abrahmson, L.P., Eggen, D., Lanier, G.N., Swier, S.R., Kelley, R.S. and Augur, M. (1986). Monitoring spruce budworm (Lepidoptera: Tortricidae) populations with pheromone-baited traps, *Env. Ent.* **15**, 152–165.

Allison, D. and Pike, K.S. (1988). An inexpensive suction trap and its use in an aphid monitoring network, *J. Agric. Ent.* **5**, 103–107.

Anderson, J.R. and de Foliart, G.R. (1961). Feeding behaviour and host preferences of some black flies (Diptera: Simuliidae) in Wisconsin, *Ann. Ent. Soc. Amer.* **54**, 716–729.

Anderson, J.R., Olkowsky, W. and Hoy, J.B. (1974). The responses of Tabanid species to CO_2-baited insect flight traps in Northern California, *Pan-Pacific Ento.* **50**, 255–268.

Angerilli, M. and McLean, J.A. (1984). Wind tunnel and field observations on western spruce budworm response to pheromone-baited traps, *J. Ent. Soc. Brit. Columbia.* **81**, 10–16.

Asman, S.M., Nelson, R.L., McDonald, P.T. *et al.* (1979). Pilot release of a sex-linked multiple translocation into a *Culex tarsalis* field population in Kern County, California, *Mosq. News.* **39**(2), 248–258.

Baker, T.C. and Roelofs, W.L. (1981). Initiation and termination of the oriental fruit moth male response to pheromone concentrates in the field, *Env. Ent.* **10**, 211–218.

Ball, S.G. (1983). A comparison made of the Diptera caught in Manitoba traps with those caught from cattle and other parts of the field ecosystem in northern England, *Bull. Ent. Res.* **73**, 527–537.

Ball, S.G. and Luff, M.L. (1981). Attractiveness of Manitoba traps to the headfly,

Hydrotaea irritans (Fallen) (Diptera: Muscidae); the effect of short-term weather fluctuations, carbon dioxide and target temperature and size, *Bull. Ent. Res.* **71**, 599–606.

Basset, I. (1988). A composite interception trap for sampling arthropods in tree canopies, *J. Aust. Ent. Soc.* **27**, 213–219.

Beier, J.C., Berry, W.J. and Craig, G.B. (1982). Horizontal distribution of adult *Aedes triseriatus* (Diptera: Culicidae) in relation to habitat structure, oviposition, and other mosquito species, *J. Med. Ent.* **19**, 239–247.

Bellas, T.E., Whittle, C.P., Rumbo, E.R. *et al.* (1988). Sex pheromones of horticultural pests. CSIRO. Biennial Report, 1985–1987. Ent. Rep. 3.7. Canberra, Australia

Bennett, G.F., Fallis, A.M. and Campbell, A.G. (1972). The response of *Simulium (Eusimulium) euryadminiculum* Davies (Diptera: Simuliidae) to some olfactory and visual stimuli, *Canad. J. Zool.* **50**, 793–800.

Bergh, J.C., Everleigh, E.S. and Seabrook, W.D. (1988). The mating status of field-collected male spruce budworm, *Choristoneura fumiferana* (Clem) (Lepidoptera: Tortricidae), in relation to trap location, sampling method, sampling date, and adult emergence, *Canad. Ent.* **180**, 821–830.

Berlyn, A.D. (1978a). The flight activity of the sheep headfly, *Hydrotaea irritans* (Fallen) (Diptera: Muscidae), *Bull. Ent. Res.* **68**, 219–228.

Berlyn, A.D. (1978b). The field biology of the adult sheep headfly, *Hydrotaea irritans* (Fallen) (Diptera: Muscidae) in south-western Scotland, *Bull. Ent. Res.* **68**, 431–436.

Berlyn, A.D. (1978c). Factors attracting the sheep headfly, *Hydrotaea irritans* (Fallen) (Diptera: Muscidae), with a note on the evaluation of repellents, *Bull. Ent. Res.* **68**, 583–588.

Beroza, M., Green, N., Gertler, S.I., Steiner, L.F. and Miyashita, D.M. (1961). New attractants for the Mediterranean fruit fly, *Agric. and Food Chem.* **9**(5), 361–365.

Bidlingmayer, W.L. (1964). The effect of moonlight on the flight activity of mosquitoes, *Ecology* **45**(1), 87–94.

Bidlingmayer, W.L. (1967). A comparison of trapping methods for adult mosquitoes; species response and environmental influence, *J. Med. Ent.* **4**(2), 200–220.

Bidlingmayer, W.L. (1971). Mosquito flight paths in relation to the environment. I. Illumination levels, orientation, and resting areas, *Ann. Ent. Soc. Amer.* **64**(5), 1121–1131.

Bidlingmayer, W.L. (1974). The influence of environmental factors and physiological stage on flight patterns of mosquitoes taken in the vehicle aspirator and truck, suction, bait and New Jersey light traps, *J. Med. Ent.* **11**(2), 119–146.

Bidlingmayer, W.L. (1975). Mosquito flight paths in relation to the environment. Effect of vertical and horizontal visual barriers. *Ann. Ent. Soc. Amer.* **68**(1), 51–57.

Bidlingmayer, W.L. (1985). The measurement of adult mosquito population changes — some considerations, *J. Amer. Mosq. Control Assn.* **1**(3), 328–348.

Bidlingmayer, W.L. and Evans. D.G. (1987). The distribution of female mosquitoes about a flight barrier, *J. Amer. Mosq. Control Assn.* **3**(3), 369–377.

Bidlingmayer, W.L., Evans, D.G. and Hansen, C.H. (1985). Preliminary study on the effects of wind velocity upon suction trap catches of mosquitoes, *J. Med. Ent.* **22**, 295–302.

Bidlingmayer, W.L. and Hem, D.G. (1979). Mosquito (Diptera: Culicidae) flight behaviour near conspicuous objects, *Bull. Ent. Res.* **69**, 691–700.

Bidlingmayer, W.L. and Hem, D.G. (1980). The range of visual attraction and the effect of competitive visual attractants upon mosquito (Diptera: Culicidae) flight, *Bull. Ent. Res.* **70**, 321–342.

Bidlingmayer, W.L. and Hem, D.G. (1981). Mosquito flight paths in relation to the environment. Effect of forest edge upon trap catches in the field, *Mosq. News.* **41**(1), 55–59.

Biggin, A.G., Dowse, J.E., Drake, V.A. *et al.* (1986). Moth flight above crops. Biennial Report 1983–85 Div. of Ent. CSIRO. Canberra, Australia.

Bishopp, F.C. (1916). Fly traps and their operation. US Dept. of Agric, *Farmers Bull.* **734**, 1–13.

Bjostad, L.D. and Roelofs, W.L. (1983). Sex pheromone biosynthesis in *Trichoplusia ni*: Key steps involve delta-ll desaturation and chain shortening, *Science* **220**, 1387–1389.

Bowden, J. (1973). The influence of moonlight on catches of insects in light traps in Africa, Part I. The moon and moonlight, *Bull. Ent. Res.* **63**, 113–128.

Bowden, J. (1981). The relationship between light-and suction-trap catches of *Chrysoperla carnea* (Stephens) (Neuroptera: Chrysopidae), and the adjustment of light-trap catches to allow for variations in moonlight, *Bull. Ent. Res.* **71**, 621–629.

Bowden, J. (1982). An analysis of factors affecting catches of insects in light traps, *Bull. Ent. Res.* **72**, 535–556.

Bowden, J. (1984). Longitudinal and seasonal changes of nocturnal illumination with a hypothesis about their effect on catches of insects in light traps, *Bull. Ent. Res.* **74**, 279–298.

Bowden, J. and Church, B.M. (1973). The influence of moonlight on catches of insects in light traps in Africa. Part II. The effect of moon phase on light trap catches, *Bull. Ent. Res.* **63**, 129–142.

Bowden, J. and Gibbs, D.G. (1973). Light-trap and suction-trap catches of insects in the northern Gezira, Sudan, in the season of southward movement of the inter-tropical front, *Bull. Ent. Res.* **62**, 571–596.

Bowden, J. and Morris, M.G. (1975). The influence of moonlight on catches of insects in light traps in Africa. III. The effective radius of a mercury vapour light and the analysis of catches using effective radius, *Bull. Ent. Res.* **65**, 303–348.

Bracken, G.K., Hanes, W. and Thorsteinson, A.J. (1962). The orientation of horse flies and deer flies (Tabanidae: Diptera). II. The role of some visual factors in the attractiveness of decay silhouettes, *Can. J. Zool.* **40**, 689–695.

Brady, J. and Shereni, W. (1988). Landing responses of the tsetse fly *Glossina morsitans morsitans* Westwood and the stable fly *Stomoxys calcitrans* (L.) (Diptera Glossinidae and Muscidae) to black-and-white patterns: a laboratory study, *Bull. Ent. Res.* **78**, 301–311.

Broadbent, L. (1948). Aphis migration and the efficiency of the trapping method, *Ann. Appl. Biol.* **35**, 379–394.

Broadbent, L., Doncaster, J.P., Hull, R. and Watson, M.A. (1948). Equipment used for trapping and identifying alate aphids. *Proc. R. Ent. Soc. Lond.* **23**, 57–58.

Broce, A.B. (1988). An improved alsynite trap for stable flies, *Stomoxys calcitrans* (Diptera: Muscidae), *J. Med. Ent.* **25**(5), 406–409.

Broce, A.B., Goodenough, J.L. and Coppedge, J.R. (1977). A wind oriented trap for screwworm flies, *J. Econ. Ent.* **70**, 413–419.

Brown, E.S., Betts, E. and Rainey, R.C. (1969). Seasonal changes in distribution of the African armyworm, *Spodoptera exempta* (Wlk) (Lepidoptera: Noctuidae) with special reference to eastern Africa. *Bull. Ent. Res.* **58**, 661–728.

Brown, E.S. and Taylor, L.R. (1971). Lunar cycles in the distribution and abundance of airborne insects in the equatorial highlands of East Africa, *J. Anim. Ecol.* **40**, 767–779.

Browne, S.M. and Bennett, G.F. (1981). Response of mosquitoes (Diptera: Culicidae) to visual stimuli. *J. Med. Ent.* **18**(6), 505–521.

Brust, R.A. (1980). Dispersal behaviour of adult *Aedes sticticus* and *Aedes vexans* (Diptera: Culicidae) in Manitoba, *Can. Ent.* **112**, 31–42.

Bursell, E. (1984). Effect of host odour on the behaviour of tsetse, *Insect. Sci. Applic*, **5**, 345–349.

Bursell, E. (1987). The effect of wind-borne odours on the direction of flight in tsetse flies, *Glossina* spp., *Physio. Ent.* **12**, 149–156.

Bursell, E., Gough, A.J.E., Beevor, P.S. *et al.* (1988). Identification of components of cattle urine attractive to tsetse flies, *Glossina* spp. (Diptera: Glossinidae), *Bull. Ent. Res.* **78**, 281–291.

Canaday, C.L. (1987). Comparison of insect fauna captured in six different trap types in a Douglas-fir forest, *Canad. Ent.* **119**, 1101–1108.

Cardé, R.T. (1979). Behavioural responses of moths to female-produced pheromones and the utilization of attractant-baited traps for population monitoring. *In* 'Movement of Highly Mobile Insects: Concepts and Methodology in Research'. (R.L. Rabb, and G.G. Kennedy eds), North Carolina State Univ. 456 pp.

Cardé, R.T. (1984). Chemo orientation in flying insects. *In* 'Chemical Ecology of Insects' (W.J. Bell and R.T. Cardé, eds), Chapman & Hall Ltd. London, pp. 111–124

Cardé, R.T. and Baker, T.C. (1984). Sexual communication with pheromones. *In* 'Chemical Ecology of Insects' W.J. Bell and R.T. Cardé, eds), Chapman and Hall Ltd, London. pp. 355–383.

Cardé, R.T. and Charlton, R.E. (1984). Olfactory sexual communication in Lepidoptera. Strategy, sensitivity and selectivity. *In* 'Insect Communication', Roy. Ent. Soc. Lond, pp. 241–265.

Cardé, R.T., Kochansky, J., Stimmel, J.F. *et al.* (1975). Sex pheromones of the European corn borer (*Ostrinia nubilalis*): cis- and trans- responding males in Pennsylvania. *Env. Ent.* **4**, 413–414.

Challier, A. and Laveissiere, C. (1973). Un noveau piege pour la capture des glossines (Glossina; Diptera, Muscidae): description et essais sut le terrain, *Cahiers ORSTOM ser Ent. Med. et Parasit.* **XI**(4), 351–362.

Chaniotis, B.N., Neely, J.M., Correa, M.A. *et al.* (1971). Natural population dynamics of phlebotomine sandflies in Panama, *J. Med. Ent.* **8**, 339–352.

Chenier, J.V.R. and Philogene, B.J.R. (1989). Evaluation of three trap designs for the capture of conifer-feeding beetles and other forest coleoptera, *Canad. Ent.* **121**, 159–167.

Collins, C.W. and Potts, S.F. (1932). Attractants for the flying gypsy moths as an aid in locating new infestations, *USDA Tech. Bull,* **336**, 43 pp.

Cooksey, L.M. and Wright, R.E. (1987). Flight range and dispersal activity of the host-seeking horse fly, *Tabanus abactor* (Diptera: Tabanidae) in north central Oklahoma, *Env. Ent.* **16**, 211–217.

Coppedge, J.R., Ahrens, E.H. and Snow, J.W. (1978). Swormlure-2 baited traps for detection of native screwworm flies, *J. Econ. Ent.* **71**, 573–575.

Critchley, E.R. *et al.* (1983). Control of pink bollworm, *Pectinophora gossypiella* (Saunders) (Lepidoptera: Gelechiidae) in Egypt by mating disruption using an aerially applied microencapsulated pheromone formulation, *Bull. Ent. Res.* **73**, 289–299.

Cytrynowicz, M., Morgante, J.S. and De Souza, H.M.L. (1982). Visual responses of South American fruit flies, *Anastrepha fraterculus*, and Mediterranean fruit flies, *Ceratitis capitata*, to coloured rectangles and spheres, *Environ. Entomol.* **11**, 1202–1210.

Cunningham, R.T. and Steiner, L.F. (1972). Field trial of cue-lure + naled on saturated fibreboard blocks for control of the melon fly by male-annihilation technique, *J. Econ. Ent.* **65**, 505–507.

Coppedge, J.R., Ahrens, E., Goodenough, J.L., Guillot, F.S. and Snow, J.W. (1977). Field comparison of liver and a new chemical mixture as attractant for the screwworm fly, *Env. Ent.* **6**, 66–68.

Danthanarayana, W. (1976). Diel and lunar flight periodicities in the light brown apple moth, *Epiphyas postvittana* (Walker) (Tortricidae) and their possible adaptive significance, *Aust. J. Zool.* **24**, 65–73.

Danthanarayana, W. (1986). Lunar periodicity of insect flight. *In* 'Insect Flight: Dispersal and Migration' (W. Danthanarayana, ed.), pp. 88–119, Springer-Verlag, Berlin.

Darling, D.C. and Packer, L. (1988). Effectiveness of Malaise traps in collecting Hymenoptera: the influence of trap design, mesh size and location, *Canad. Ent.* **120**, 787–796.

David, C.T., Kennedy, J.S., Ludlow, A.R. *et al.* (1982). A reappraisal of insect flight towards a distant point source of wind-borne odor, *J. Chem. Ecol.* **8**(9), 1207–1215.

David, C.T., Kennedy, J.S. and Ludlow, A.R. (1983). Finding of a sex pheromone source by gypsy moths released in the field, *Nature* **303**, 30th June, 804–806.

Davies, J.B. (1975). Moonlight and the biting activity of *Culex(Melanoconion) portesi*, Senevet & Abonnenc, and *C.(M) taeniopus*, D & K (Diptera, Culicidae) in Trinidad forests, *Bull. Ent. Res.* **68**, 81–96.

Davies, J.B. (1978). Attraction of *Culex portesi* Senevet & Abonnenc and *Culex taeniopus* Dyar & Knab (Diptera: Culicidae) to 20 animal species exposed in a Trinidad forest, *Bull. Ent. Res.* **68**, 707–719.

Davies, L. and Roberts, D.M. (1973). A net and a catch-segregating apparatus mounted in a motor vehicle for field studies on flight activity of Simuliidae and other insects, *Bull. Ent. Res.* **63**, 103–112.

Davies, L. and Williams, C.B. (1962). Studies on black flies (Diptera: Simuliidae) taken in a light trap in Scotland. I. Seasonal distribution, sex ratio and internal condition of catches, *Trans. Roy. Ent. Soc. Lond.* **114**, 1–20.

Deansfield, R.D., Brightwell, R., Onah, J. and Okolo, C.J. (1982). Population dynamics of *Glossina morsitans submorsitans*, Newstead, and *G. tachinoides* Westwood (Diptera: Glossinidae) in sub-Sudan savanna in northern Nigeria. I. Sampling methodology for adults and seasonal changes in numbers caught in different vegetation types, *Bull. Ent. Res.* **72**, 175–192.

Dent, D.R. and Pawar, C.S. (1988). The influence of moonlight and weather on catches of *Helicoverpa armigera* (Hubner) (Lepidoptera: Noctuidae) in light and pheromone traps, *Bull. Ent. Res.* **78**, 365–377.

Disney, R.H.L. (1972). Observations on chicken-biting blackflies in Cameroon with a discussion of parous rates of *Simulium damnosum*, *Ann. Trop. Med. Parasit.*, **66**(1), 149–158.

Douthwaite, R.J. (1978). Some effects of weather and moonlight on light-trap catches of the armyworm *Spodoptera exempta* (Walker) (Lepidoptera, Noctuidae) at Muguga, Kenya, *Bull. Ent. Res.* **68**, 533–542.

Drew, R.A.I. (1974). The responses of fruit fly species (Diptera: Tephritidae) in the south Pacific area to male attractants, *J. Aust. Ent. Soc.* **13**, 267–270.

Drew, R.A.I. and Hooper, G.H.S. (1981). The responses of fruit fly species (Diptera: Tephritidae) in Australia to various attractants, *J. Aust. Ent. Soc.* **20**, 201–205.

Drummond, F., Groden, E. and Prokopy, R.J. (1984). Comparative efficacy and optimal positioning of traps for monitoring apple maggot flies (Diptera: Tephritidae). *Env. Ent.* **13**, 232–235.

Easton, E.R. (1987). Mosquito surveillance employing New Jersey light traps on Indian reservations in Iowa, Nebraska and South Dakota in 1984 and 1985, *J. Amer. Mosq. Control Assn.* **3**(1), 70–73.

El Bashir, S., El Jack, M.H. and El Hadi, H.M. (1976). The diurnal activity of the chicken-biting black fly, *Simulium griseicolle* Becker (Diptera: Simuliidae) in northern Sudan, *Bull. Ent. Res.* **66**, 481–487.

Elkinton, J.S. and Cardé, R.T. (1988). Effects of intertrap distance and wind direction on the interaction of gypsy moth (Lepidoptera: Lymantriidae) pheromone-baited traps, *Env. Ent.* **17**, 764–769.

Elkinton, J.S., Schal, C., Ono, T. and Cardé, R.T. (1987). Pheromone puff trajectory upwind flight of male gypsy moth in the forest, *Physio. Ent.* **12**, 399–406.

Economopoulos, A.P. (1977). Controlling *Dacus oleae* by fluorescent yellow traps, *Ent. Expl. Appl.* **22**, 183–190.

Economopoulos, A.P. Attraction of *Dacus oleae* (Gmelin) (Diptera, Tephritidae) to odour and colour traps, *Z. ang. Ent.* **88**, 90–97.

Farrow, R.A. (1974). A modified light trap for obtaining large samples of night-flying locusts and grasshoppers, *J. Aust. Ent. Soc.* **13**, 357–360.

Farrow, R.A. (1977). First captures of the migratory locust, *Locusta migratoria* L., at light traps and their ecological significance, *J. Austral. Ent. Soc.* **16**, 59–61.

Farrow, R.A. (1984). Detection of transocean migration of insects to a remote island in the Coral sea., Willis Island, *Aust. J. Ecol.* **9**, 253–272.

Farrow, R.A. and Daly, J.C. (1987). Long range movements as an adaptive strategy in the genus *Heliothis* (Lepidoptera, Noctuidae): A review of its occurrence and detection in four pest species, *Aust. J. Zool.* **35**, 1–24.

Farrow, R.A. and Dowse, J.E. (1984). Method of using kites to carry tow nets to the upper air for sampling migratory insects and its application to radar entomology, *Bull. Ent. Res.* **74**, 87–95.

Farrow, R.A., Drake, V.A. and Harris, P.L. (1988). Development of techniques for studying the migration and light behaviour of insects, CSIRO Biennial Report 1985–87, Canberra, Australia 5.4–5.5.

Fein, B.L., Reissig, W.H. and Roelofs, W.L. (1982). Identification of apple volatiles attractive to the apple maggot, *Rhagoletis pomonella*, *J. Chem. Ecol.* **8**, 1473–1487.

Finch, S. and Collier, R.N. (1989). Diptera caught on sticky boards in certain vegetable crops, *Ent. Expt. et Appl.* **52**(1), 23–27.

Fitt, G.P. and Van den Elst, G. (1988). The usefulness of pheromone traps as indicators of *Heliothis* activity in cotton, CSIRO. Biennial Report 1985–1987. 5, 11. Canberra, Australia.

Fitt, G.P., Zalucki, M.P. and Twine, P. (1989). Temporal and spatial pattern of pheromone-trap catches of *Helicoverpa* spp. (Lepidoptera: Noctuidae) in cotton growing areas of Australia, *Bull. Ent. Res.* **79**, 145–161.

Fletcher-Howell, G., Ferro, D.N. and Butkewich, S. (1983). Pheromone and blacklight trap monitoring of adult European cornborer (Lepidoptera: Pyralidae) in western Massachussetts, *Env. Ent.* **12**, 531–534.

Flint, S. (1985). A comparison of various traps for *Glossina* spp. (Glossinidae) and other Diptera, *Bull. Ent. Res.* **75**, 529–534.

Fredeen, F.J.H. (1961). A trap for studying the attacking behaviour of black flies, *Simulium arcticum* Mall, *Canad. Ent.* **93**, 73–78.

Gaydecki, P.A. (1984). A quantification of the behavioural dynamics of certain Lepidoptera in response to light. Ph.D. thesis. Cranfield Institute of Technology. Ecological Physics Research Group, pp. 157 plus appendices.

Gillespie, D.R. and Quiring, R. (1987). Yellow sticky traps for detecting and monitoring greenhouse whitefly (Homoptera: Aleyrodidae) adults on greenhouse tomato crops, *J. Econ. Ent.* **80**, 675–679.

Gillies, M.T. (1969). The ramp-trap, an unbaited device for flight studies of mosquitoes, *Mosq. News.* **29**(2), 189–193.

Gillies, M.T. (1980). The role of carbon dioxide in host finding by mosquitoes (Diptera: Culicidae), *Bull. Ent. Res.* **70**, 525–532.

Gillies, M.T., Jones, M.D.R. and Wilkes, T.J. (1978). Evaluation of a new technique for recording the direction of flight of mosquitoes (Diptera: Culicidae) in the field, *Bull. Ent. Res.* **68**, 145–152.

Gillies, M.T. and Wilkes, T.J. (1969). A comparison of the range of attraction of animal baits and of carbon dioxide for some West African mosquitoes, *Bull. Ent. Res.* **59**, 441–456.

Gillies, M.T. and Wilkes, T.J. (1970). The range of attraction of single baits for some West African mosquitoes, *Bull. Ent. Res.* **60**, 225–235.

Gillies, M.T. and Wilkes, T.J. (1972). The range of attraction of animal baits and carbon dioxide for mosquitoes. Studies in a freshwater area of West Africa, *Bull. Ent. Res.* **61**, 389–404.

Gillies, M.T. and Wilkes, T.J. (1976). The vertical distribution of some West African mosquitoes (Diptera: Culicidae) over open farmland in a freshwater area of the Gambia, *Bull. Ent. Res.* **66**, 5–15.

Gillies, M.T. and Wilkes, T.J. (1978). The effect of high fences on the dispersal of some West African mosquitoes (Diptera: Culicidae), *Bull. Ent. Res.* **68**, 401–408.

Glover, T.J., Tang, X.H. and Roelofs, W.L. (1987). Sex pheromone blend discrimination by male moths from E and Z strains of European cornborer, *J. Chem. Ecol.* **13**, 143–151.

Goodwin, S. and Danthanarayana, W. (1984). Flight activity of *Plutella xylostella* (L) (Lepidoptera: Yponomeutidae), *J. Aust. Ent. Soc.* **23**, 235–240.

Green, C.H. (1989). The use of two-coloured screens for catching *Glossina palpalis (Robineau-Desvoidy) (Diptera: Glossinidae)*, *Bull. Ent. Res.* **79**, 81–93.

Green, C.H. and Flint, S. (1986). Analysis of colour effects in the performance of the F2 trap against *Glossina pallidipes* Austen, and *G.morsitans morsitans* Westwood (Diptera: Glossinidae), *Bull. Ent. Res.* **76**, 409–418.

Greenbank, D.O., Schaefer, G.W. and Rainey, R.C. (1980). Spruce budworm (Lepidoptera: Tortricidae) moth flights and dispersal: new understanding from canopy observations, radar and aircraft, *Mem. Ent. Soc. Canad.* **110**, p. 49.

Gressitt, J.L., Sedlacek, J., Wise, K.A.J. and Yoshimoto, C.M. (1961). A high speed airplane trap for air-borne organisms, *Pacific Insects.* **3**, 549–555.

Hall, D.R., Beevor, P.S., Cork, A. *et al.* (1984). 1-Octen-3-ol; a potent olfactory stimulant and attractant for tsetse isolated from cattle odours, *Insect. Sci. Applic.* **5**, 335–339.

Hanec, W. and Bracken, G.K. (1964). Seasonal and geographic distribution of Tabanidae (Diptera) in Manitoba, based on females captured in traps, *Canad. Ent.* **86**, 136–396.

Haniotakis, G.E., Kozyrakis, E. and Bonatsos, C. (1986). Control of the olive fruit fly, *Dacus oleae* Gmel. (Dipt., Tephritidae) by mass trapping; Pilot scheme feasibility study, *J. Appl. Ent.* **101**, 343–352.

Hansens, E.J., Bosler, E.M. and Robinson, J.W. (1971). Use of traps for study and control of saltmarsh greenhead flies, *J. Econ. Ent.* **64**(6), 1481–1486.

Hardwick, D.F. (1972). The influence of temperature and moon phase on the activity of noctuid moths, *Canad. Ent.* **164**, 1767–1770.

Hargrove, J.W. (1976). The effect of human presence on the behaviour of tsetse (*Glossina* spp.) (Diptera: Glossinidae) near a stationary ox. *Bull. Ent. Res.* **66**, 173–178.

Hargrove, J.W. (1977). Some advances in the trapping of tsetse (*Glossina* spp.)

and other flies, *Ecol. Ent.* **2**, 123–137.

Hargrove, J.W. and Vale, G.A. (1978). The effect of host odour concentration on catches of tsetse flies (Glossinidae) and other Diptera in the field, *Bull. Ent. Res.* **68**, 607–612.

Harris, E.J. and Lee, C.Y.L. (1987). Seasonal and annual distribution of the Mediterranean fruit fly (Diptera: Tephritidae) in Honolulu and suburban areas of Oahu, Hawaii, *Env. Ent.* **16**(6), 1273–1282.

Hart, W.D., Meyerdirk, M., Sanchez, W. and Rhode, R. (1978). Development of a trap for the citrus blackfly, *Aleurocanthus woglumi* Ashby, *S. West Ent.* **3**, 219–225.

Hartstack, A.W., Hollingsworth, J.P. and Lindquist, D.A. (1968). A technique for measuring trapping efficiency of electric insect traps, *J. Econ. Ent.* **61**, 546–552.

Hill, A.R. (1986). Reduction in trap capture of female fruit flies (Diptera: Tephritidae) when synthetic male lures are added, *J. Aust. Ent. Soc.* **25**, 211–214.

Hill, A.R. and Hooper, G.H.S. (1984). Attractiveness of various colours to Australian tephritid fruit flies in the field, *Entomol. Expt. Appl.* **35**, 119–128.

Hines, J.W. and Heikkenen, H.J. (1977). Beetles attracted to severed Virginia pine (*Pinus virginiana* Mill), *Env. Ent.* **6**, 123–127.

Houseweart, M.W., Jennings, D.T. and Sanders, C.J. (1981). Variables associated with pheromone traps for monitoring spruce budworm populations (Lepidoptera: Tortricidae), *Canad. Ent.* **113**, 527–537.

Howell, J.F. (1972). An improved sex attractant trap for codling moths, *J. Econ. Ent.* **65**, 609–611.

Howell, J.F. (1984). A new pheromone trap for monitoring codling moth (Lepidoptera: Olethreutidae) populations, *J. Econ. Ent.* **77**, 1612–1614.

Hsiao, H.S. (1973). Flight paths of night-flying moths to light, *J. Insect Physio.* **19**, 1971–1976.

Hughes, R.D. (1970). The seasonal distribution of bushfly (*Musca vetustissima*, Walker) in south-east Australia, *J. Anim. Ecol.* **39**, 691–706.

Hughes, R.D. (1974). Variation in the proportion of different reproductive stages of female bushflies (*Musca vetustissima*, Wlk) (Diptera; Muscidae) in bait catches as a cause of error in population estimates, *Bull. Ent. Res.* **64**, 65–71.

Hughes, R.D., Duncan, P. and Dawson, J. (1981). Interactions between Camargue horses and horseflies (Diptera: Tabanidae), *Bull. Ent. Res.* **71**, 227–242.

Johnson, C.G. (1950). The comparison of suction trap, sticky trap and townet for the quantitative sampling of small airborne insects, *Ann. Appl. Biol.* **37**, 268–285.

Johnson, C.G. (1957). The distribution of insects in the air and the empirical relation of density to height, *J. Anim. Ecol.* **26**, 479–494.

Johnson, C.G. (1969). 'Migration and Dispersal of Insects by Flight', Methuen, London.

Johnson, C.G. and Taylor, L.R. (1955). The development of large suction traps for airborne insects, *Ann. App. Biol.* **43**, 51–62.

Johnson, C.G., Crosskey, R.W. and Davies, J.B. (1982). Species composition and cyclical changes in numbers of savanna blackflies (Diptera: Simuliidae) caught by suction traps in the Onchocerciasis Control Programme area in W. Africa, *Bull. Ent. Res.* **72**, 39–63.

Jones, C.M., Oehler, D.D., Snow, J.W. and Grabbe, R.R. (1976). A chemical attractant for screwworm flies, *J. Econ. Ent.* **69**, 389–391.

Jones, R.H. (1961). Some observations on biting flies attacking sheep, *Mosq. News.* **21**, 113–115.

Jones, V.P. (1988). Longevity of apple maggot (Diptera: Tephritidae) lures under laboratory and field conditions in Utah, *Env. Ent.* **17**(4), 704–708.

Jupp, P.G. (1978). A trap to collect mosquitoes attracted to monkeys and baboons, *Mosq. News.* **38**, 288–289.

Katsoyannos, B.L. (1987). Effect of colour properties of spheres on their attractiveness for *Ceratitis capitata* (Wiedmann) in the field. *J. Appl. Ent.* **104**, 79–85.

Katsoyannos, B.L., Panagiotidou, K. and Kechagia, I. (1986). Effect of colour properties on the selection of oviposition site by *Ceratitis capitata*, *Entomol. Exp. Appl*, **42**, 187–193.

Kehat, M., Gothilf, S., Dunkelblum, E. and Greenberg, S. (1982). Sex pheromone traps as a means of improving control programmes for the cotton bollworm, *Heliothis armiger* (Lepidoptera: Noctuidae), *Env. Ent.* **11**, 727–729.

Kendall, D.M., Jennings, D.T. and Houseweart, M.W. (1982). A large capacity pheromone trap for spruce budworm moths (Lepidoptera: Tortricidae), *Canad. Ent.* **114**, 461–463.

Kennedy, G.G. and Anderson, T.E. (1980). European cornborer (Pyralidae) trapping in North Carolina with various sex pheromone component blends, *J. Econ. Ent.* **73**, 642–646.

Knight, A.L. and Croft, B.A. (1987). Temporal patterns of competition between a pheromone trap and caged female moths for males of *Argyrotaenia citrana* (Lepidoptera: Tortricidae) in a semi-enclosed courtyard, *Env. Ent.* **15**, 1185–1192.

Koch, K. and Spielberger, U. (1979). Comparison of hand nets, biconical traps and an electric trap for sampling *Glossina palpalis palpalis* (Robineau-Desvoidy) and *G. tachinoides* Westwood (Diptera: Glossinidae) in Nigeria, *Bull. Ent. Res.* **69**, 243–253.

Kring, J.B. (1970). Red spheres and yellow panels combined to attract apple maggot flies, *J. Econ. Ent.* **63**(2), 466–469.

Kundu, H.L. (1985). A note on problems and prospects of mechanical traps for pest/vector research and control. *In* 'Use of Traps for Pest/Vector Control'. Proc. Nat. Seminar. Mohanpur, West Bengal, pp. 119–132.

Laveissiere, C., Couret, D. and Challier, A. (1979). Description and design details of a biconical trap used in the control of tsetse flies along the banks of rivers and streams WHO/VBC/79.746. WHO documentary series, Geneva.

Lewis, T. and Macauley, E.D.M. (1976). Design and elevation of sex-attractant traps for pea moth, *Cydia nigricana* (Steph) and the effect of plume shape on catches, *Ecol. Ent.* **1**, 175–187.

Lewis, T., Wall, C., Macauley, E.D.M. and Greenway, A.R. (1975). The behavioural basis of a pheromone monitoring system for pea moth, *Cydia nigricana*, *Ann. Appl. Biol.* **80**, 257–274.

Lingren, P.D., Henneberry, T.J. and Bariola, L.A. (1980). Nocturnal behaviour of adult cotton leafperforators in cotton, *Ann. Ent. Soc. Am.* **73**, 44–48.

Lingren, P.D., Sparks, A.N., Raulston, J.R. and Wolf, W.W. (1978). Applications for nocturnal studies of insects, *Bull. Ent. Soc. Am.* **24**, 206–212.

Linn, C.E., Bjostad, L.B., Du, J.W. and Roelofs, W.L. (1984). Redundancy in a chemical signal: Behavioural responses of male *Trichoplusia ni* to a 6-component sex pheromone blend, *J. Chem. Ecol* **10**, 1635–1658.

Linn, C.E., Campbell, M.G. and Roelofs, W.L. (1988a). Temperature modulation of behavioural thresholds controlling male moth sex pheromone response selectivity, *Physio. Ent.* **13**, 59–67.

Linn, C.E., Hammond, A., Du, J. and Roelofs, W.L. (1988b). Specificity of male response to multi-component pheromones in noctuid moths *Trichoplusia ni* and *Pseudoplusia includens*, *J. Chem. Ecol.* **14**, 47–57.

Macauley, E.D.M., Tatchell, G.M. and Taylor, R. (1988). The Rothamsted Insect Survey '12 metre' suction trap, *Bull. Ent. Res.* **78**, 121–129.

McCreadie, J.W., Colbo, M.H. & Bennett, G.F. (1984). A trap design for the collection of haematophagous Diptera from cattle, *Mosq. News.* **44**, 212–216.

McGeachie, W.J. (1987). The effect of air temperature, wind vectors and nocturnal illumination on the behaviour of moths at mercury vapour light traps. Ph.D. thesis. Ecological Physics Research Group, Cranfield Institute of Technology. 81 pp plus appendices.

McGeachie, W.J. (1988). A remote sensing method for the estimation of light trap efficiency, *Bull. Ent. Res.* **78**, 379–385.

McLeod, D.G.R. and Starratt, A.N. (1978). Some factors influencing pheromone trap catches of the European cornborer, *Ostrinia nubialis* (Lepidoptera: Pyralidae), *Canad. Ent.* **110**, 51–55.

Mackley, J.W. and Brown, H.E. (1984). Swormlure-4: a new formulation of the Swormlure-2 mixture as an attractant for adult screwworm flies *Cochliomyia hominivorax* (Diptera: Calliphoridae), *J. Econ. Ent.* **77**, 1264–1268.

McNally, P.S. and Barnes, M.M. (1981). Effect of codling moth pheromone trap placement, orientation and density, on trap catches, *Env. Ent.* **10**, 22–26.

McPhail, M. (1937). Relation of time of day, temperature and evaporation to attractiveness of fermenting sugar solution to Mexican fruit fly, *J. Econ. Ent.* **30**, 793–799.

Mahrt, G.G., Stoltz, R.L., Blickenstaff, C.C. and Holtzer, T.O. (1987). Comparisons between blacklight and pheromone traps for monitoring the western bean cutworm (Lepidoptera: Noctuidae) in south central Idaho, *J. Econ. Ent.* **80**, 242–247.

Mason, P.G. (1986). Evaluation of a 'cow-type' silhoutte trap with and without CO_2 bait for monitoring populations of *Simulium luggeri* (Diptera: Simuliidae), *J. Amer. Mosq. Control Assn.* **2**(4), 482–484.

Meyerdirk, D.E., Hart, W.G. & Burnside, J. (1979). Evaluation of a trap for the citrus blackfly *Aleurocanthus woglumi* (Homoptera: Aleyrodidae), *Canad. Ent.* **111**, 1127–1129.

Meyerdirk, D.E. and Oldfield, G.N. (1985). Evaluation of trap colour and height placement for monitoring *Circulifer tenellus* (Baker) (Homoptera: Cicadellidae), *Canad. Ent.* **117**, 505–511.

Miller, C.A. (1971). The spruce budworm in eastern N. America. *Proc. 3rd Tall Timbers Conf. Ecol. Animal Control Habitat Management* 169–177.

Miller, C.K. and McDougall, G.A. (1973). Spruce budworm moth trapping using virgin females, *Canad. J. Zool.* **51**, 853–858.

Miller, J.R. and Roelofs, W.L. (1978). Sustained flight tunnel for measuring insect responses to wind-borne sex pheromones, *J. Chem. Ecol.* **4**, 142–149.

Mitchell, C.J., Darsie, R.F., Monath, T.P. *et al.* (1985a). The use of an animal baited net trap for collecting mosquitoes during western encephalitis investigations in Argentina, *J. Amer. Mosq. Control Assn.* **1**(1), 43–47.

Mitchell, C.J., Monath, T.P., Sabattini, M.S. *et al.* (1985b). Arbovirus investigations in Argentina II. Arthropod collections and virus isolations from mosquitoes, *Amer. J. Trop. Med. Hyg.* **34**, 945–955.

Morris, K.R.S. (1960). Trapping as a means of studying the game tse-tse *Glossina pallidipes*, *Aust. Bull. Ent. Res.* **51**, 533–557.

Morris, K.R.S. (1963). A study of African tabanids made by trapping, *Acta. Tropica* **20**(1), 16–34.

Morris, R.F. (1955). The development of sampling techniques for forest insect defoliators, with particular reference to the spruce budworm, *Canad. J. Zool.* **33**(4), 225–294.

Morris, K.R.S. and Morris, M.G. (1949). The use of traps against tsetse in West Africa, *Bull. Ent. Res.* **39**, 491–528.

Morton, R., Tuart, L.D. and Wardhaugh, K.G. (1981). The analysis and standardization of light trap catches of *Heliothis armiger* (Hubner) and *H. punctiger* (Lepidoptera, Noctuidae), *Bull. Ent. Res.* **7**, 207–225.

Moser, J.C. and Brown, L.E. (1978). A nondestructive trap for *Dendroctones frontalis*, *J. Chem. Ecol.* **4**, 1–7.

Muirhead-Thomson, R.C. (1982). Behaviour Patterns of Blood-Sucking Flies, Pergamon Press, Oxford.

Mukhopadhyay, S., Chakravarti, S. and Mukhopadhyay, S. (1985). The use of light traps for studying the biometeorological relations of the rice green leafhoppers. *In* 'Use of Traps for Pest/vector Research and Control', 91–102, Proc. Nat. Seminar. Mohanpur, West Bengal.

Mulhern, T.D. (1932) A new development in mosquito traps. *Proc. 21st Ann. Meet. New Jersey Mosq. Exter. Assn.* **137**.

Mulhern, T.D. (1942). New Jersey mechanical trap for mosquito surveys. *New Jersey Agric. Expt. Stn. Circular 421*, 1–8.

Mulhern, T.D. (1985). New Jersey Mechanical trap for mosquito surveys, *J. Amer. Mosq. Control Assn.* **1**(4), 411–418.

Muller, M.J. and Murray, M.D. (1977). Blood-sucking flies feeding on sheep in Eastern Australia, *Aust. J. Zool.* **25**, 75–85.

Muller, M.J., Murray, M.D. and Edwards, J.A. (1981). Blood sucking midges and mosquitoes feeding on mammals at Beatrice Hill, N.T., *Aust. J. Zool.* **29**, 573–588.

Nakagawa, S., Chambers, D.L., Urago, T. and Cunningham, R.T. (1971). Trap-lure combinations for surveys of Mediterranean fruit flies in Hawaii, *J. Econ. Ent.* **64**(5), 1211–1213.

Nakagawa, S., Prokopy, R.J., Wong, T.T.Y. *et al.* (1978). Visual orientation of *Ceratitis capitata* flies to model fruit, *Ent. Expt. et Appl.* **24**, 193−198.

Nath, D.K. and Banerjee, D.K. (1985). Analysis and interpretation of three years light trap catches of the green leaf hoppers at Sarul, Burdwan, with the emphasis on the method of analysis for occurrence prediction. *In*: 'Use of Traps for Pest/vector Research and Control', pp. 68−75. Proc. Nat. Seminar, Mohanpur, West Bengal.

Nelson, R.L. and Bellamy, R.E. (1971). Patterns of flight activity of *Culicoides variipennis* (Coquillet) (Diptera; Ceratopogonidae), *J. Med. Ent.* **8**(3), 283−291.

Nelson, R.L. and Milby, M.M. (1980). Dispersal and survival of field and laboratory strains of *Culex tarsalis* (Diptera: Culicidae), *J. Med. Ent.* **17**, 146−150.

Nelson, R.L., Milby, M.M., Reeves, W.C. and Fine, P.E.M. (1978). Estimates of survival, population size, and emergence of *Culex tarsalis* at an isolated site, *Ann. Ent. Soc. Amer.* **7**, 801−808.

Nemec, S.J. (1971). Effects of lunar phases on light-trap collections and populations of bollworm moths, *J. Econ. Ent.* **64**, 860−862.

Novak, R.J., Peloquin, J. and Rohrer, W. (1981). Vertical distribution of adult mosquitoes (Diptera: Culicidae) in a northern deciduous forest in Indiana, *J. Med. Ent.* **18**(2), 116−122.

Novak, M.A. and Roelofs, W.L. (1985). Behaviour of male redbanded leafroller moths *Argyrotaenia velutinana* (Lepidoptera: Tortricidae) in small disruption plots, *Env. Ent.* **14**, 12−16.

Oloumo-Sadeghi, H., Showers, W.B. and Reed, G.L. (1975). European corn borer. Lack of synchrony of attraction to sex pheromones and capture in light traps, *J. Econ. Ent.* **68**, 663−667.

Paliniswamy, P. and Underhill, E.W. (1988). Mechanisms of orientation disruption by sex pheromone components in the redbacked cutworm, *Euxoa cohrogaster* (Guenee) (Lepidoptera: Noctuidae), *Env. Ent.* **17**(3), 432−441.

Pawar, C.S., Sitanantham, S., Sharma, H.C. *et al.* (1985). Use and development of insect traps at ICRISAT *In*: 'Use of Traps for Pest/vector Research and Control, pp. 133−142, Mohanpur, West Bengal.

Perry, J.N. and Wall, C. (1984). Short term variations in catches of the pea moth *Cydia nigricana* in interacting pheromone traps, *Ent. Expt. et Appl.* **36**, 145−149.

Persson, B. (1971). Influence of light on flight activity of Noctuids (Lepidoptera) in south Sweden, *Entomol. Scand.* **2**, 215−232.

Persson, B. (1976). Influence of weather and nocturnal illumination on the activity and abundance of populations of noctuids (Lepidoptera) in south coastal Queensland, *Bull. Ent. Res.* **66**, 33−63.

Phelps, R.J. (1968). A falling cage for sampling tsetse flies (Glossina: Diptera). *Rhod. J. Agric. Res.* **6**, 47−53.

Pivnick, K.A., Barton, D.L., Millar J.G. and Underhill, E.W. (1988). Improved pheromone trap exclusion of the Bruce spanworm *Operophtera bruceata* (Holst) (Lepidoptera: Geometridae) when monitoring winter moth *Operophtera brumata* (L) populations, *Canad. Ent.* **120**, 389−396.

Prokopy, R.J., Bennet, E.W. and Bush, A.L. (1971). Mating behaviour of *Rhagoletis pomonella* (Diptera: Tephritidae). I. Site of assembly, *Canad. Ent.* **103**, 1405–1409.

Prokopy, R.J., Bennet, E.W. and Bush, G.L. (1972). Mating behaviour of *Rhagoletis pomonella* (Diptera: Tephritidae). II. Temporal organization, *Canad. Ent.* **104**, 97–104.

Provost, M.W. (1959). The influence of moonlight on light-trap catches of mosquitoes, *Ann. Ent. Soc. Amer.* **52**, 261–271.

Ramaswamy, S.B. and Cardé, R.T. (1982). Nonsaturating traps and long-life attractant lures for monitoring spruce budworm males, *J. Econ. Ent.* **75**, 126–129.

Reisen, W.K. and Pfuntner, A.R. (1987). Effectiveness of five methods for sampling adult *Culex* mosquitoes in rural and urban habitats in San Bernardino county, California, *J. Amer. Mosq. Control Assn.* **3**(4), 601–606.

Reissig, W.H. (1974). Field tests of traps and lures for the apple maggot, *J. Econ. Ent.* **67**, 484–486.

Reissig, W.H. (1975). Performance of apple maggot traps in various tree canopy positions, *J. Econ. Ent.* **64**, 534–538.

Reissig, W.H., Fein, B.L. and Roelofs, W.L. (1982). Field tests of synthetic apple volatiles as apple maggot (Diptera: Tephritidae) attractants, *Env. Ent.* **11**, 1294–1298.

Reissig, W.H., Novak, M.A. and Roelofs, W.L. (1978). Orientation disruption of *Argyrotaenia velutinana* and *Choristoneura rosaceana* male moths, *Env. Ent.* **7**, 631–632.

Reissig, W.H., Stanley, B.H., Roelofs, W.L. and Scharz, M.R. (1985). Tests of synthetic apple volatiles in traps as attractants for apple maggot flies (Diptera: Tephritidae) in commercial apple orchards, *Env. Ent.* **14**, 55–59.

Reling, D. and Taylor, R.A.J. (1984). A collapsible tow net used for sampling arthropods by airplane, *J. Econ. Ent.* **77**, 1615–1617.

Riedl, H., Croft, B.A. and Howitt, A.J. (1976). Forecasting codling moth phenology based on pheromone trap catches and physiological time models, *Canad. Ent.* **108**, 449–460.

Rioux, J.A. and Golvan, Y.J. (1969). Epidemiologie des Leishmanioses dans le sud de la France. Monograph de l'Institut National de la Sante et de la Recherche Medicale, Paris 223 pp.

Roach, S.H. (1975). *Heliothis zea* and *H. virescens*; moth activity as measured by backlight and pheromone traps, *J. Econ. Ent.* **68**, 17–21.

Roberts, R.H. (1972). The effectiveness of several types of Malaise traps for the collection of *Tabanidae* and *Culicidae*, *Mosq. News* **32**, 542–547.

Roberts, R.H. (1975). Influence of trap screen age on collections of tabanids in Malaise traps, *Mosq. News* **35**, 538–539.

Roberts, R.H. (1976a). The comparative efficiency of six trap types for the collection of *Tabanidae (Diptera)*, *Mosq. News* **36**, 530–535.

Roberts, R.H. (1976b). Attitude distribution of *Tabanidae* as determined by Malaise trap collections, *Mosq. News* **36**, 518–520.

Roelofs, W.L. (1978). Threshold hypothesis for pheromone perception, *J. Chem. Ecol.* **4**, 658–699.

Roelofs, W.L. and Cardé, R.T. (1974). Oriental fruit moth and lesser appleworm attractant mixtures refined, *Env. Ent.* **3**, 586–588.

Roelofs, W.L., Comeau, A. and Selle, R. (1969). Sex pheromone of the oriental fruit moth, *Nature*, **224**, 723.

Roelofs, W.L., Comeau, A., Hill, A. and Milicivic, C. (1971). Sex attractants of the codling moth. Characterization with electroentennogram techniques, *Science*, **174**, 277–299.

Roelofs, W.L., Cardé, R.T., Bartell, R.J. and Tierney, P.G. (1972). Sex attraction trapping of the European corn borer in New York, *Environ. Ent.* **1**, 606–608.

Rogers, D.J. and Randolph, S.E. (1978). A comparison of electric trap and handnet catches of *Glossina palpalis palpalis* (Robineau-Desvoidy) and *G. tachinoides* Westwood (Diptera: Glossinidae) in the Sudan vegetation zone of northern Nigeria, *Bull. Ent. Res.* **68**, 283–297.

Rogers, D.J. and Smith, D.T. (1977). A new electric trap for tsetse flies, *Bull. Ent. Res.* **67**, 153–159.

Rothschild, G.H.L. (1974). Problems in defining synergists and inhibitors of the oriental fruit moth pheromone by field experimentation, *Ent. Expt. et Appl.* **17**, 294–302.

Rothschild, G.H.L. and Minks, A.K. (1974). Time of activity of male oriental fruit moths at pheromone sources in the field, *Env. Ent.* **3**(6), 1003–1007.

Rothschild, G.H.L. and Minks, A.K. (1977). Some factors influencing the performance of pheromone traps for oriental fruit moth in Australia, *Ent. Exp. et Appl.* **22**, 171–182.

Rothschild, G.H.L., Vickers, R.A., Bakker, S.E. *et al.* (1988a). Mating disruption of *Heliothis*. CSIRO Biennial Report, 1985–87, 39, Canberra, Australia.

Rothschild, G.H.L., Vickers, R.A. and Bell, R.E. (1988b). Mating disruption of the current borer *Synanthedon tipuliformis*. CSIRO, Biennial Report, 1985–1987, 39, Canberra, Australia.

Rothschild, G.H.L., Vickers, R.A. and Morton, R. (1984). Monitoring the oriental fruit moth, *Cydia molesta* (Busck) (Lepidoptera: Tortricidae), with pheromone traps and bait pails in peach orchards in south-eastern Australia, *Protection Ecology* **6**, 115–136.

Samways, M.J. (1987a). Prediction of upsurges in population of the insect vector, *Trioza erytraea* (Hemiptera: Triozidae) of citrus greening disease using low cost trapping, *J. Appl. Ecol.* **24**, 881–891.

Samways, M.J. (1987b). Phototactic responses of *Trioza erytraea* (Del Guercio) (Hemiptera; Triozidae) to yellow coloured surfaces, and an attempt at commercial suppression using yellow barriers and trap trees, *Bull. Ent. Res.* **77**, 91–98.

Sanders, C.J. (1978). Evaluation of sex attractant traps for monitoring spruce budworm populations (Lepidoptera: Tortricidae), *Canad. Ent.* **110**, 43–50.

Sanders, C.J. (1984). Sex pheromones of the spruce budworm (Lepidoptera: Tortricidae); evidence for a missing component *Can. Ent.* **116**, 93–100.

Sanders, C.J. (1986). Evaluation of high-capacity, nonsaturating sex pheromone traps for monitoring population densities of spruce budworm (Lepidoptera: Tortricidae), *Canad. Ent.* **118**, 611−619.

Sanders, C.J. (1988). Monitoring spruce budworm population density with six pheromone traps, *Canad. Ent.* **120**, 175−183.

Sanders, C.J. and Meighen, E.A. (1987). Controlled-release sex pheromone lures for monitoring spruce budworm populations, *Canad. Ent.* **119**, 305−313.

Sanders, C.J. and Weatherston, J. (1976). Sex pheromone of the eastern spruce budworm. Optimum blend of trans- and cis-11-tetradecenal, *Can. Ent.* **108**, 1285−1290.

Schaefer, G.W. and Bent, G.A. (1984). An infra-red remote sensing system for the active detection and automatic determination of insect flight trajectories, *Bull. Ent. Res.* **74**, 261−278.

Schaefer, G.W., Bent, G.A. and Allsopp, K. (1985). Radar and opto-electronic measurements of the effectiveness of Rothamsted Insect Survey suction traps, *Bull. Ent. Res.* **75**, 701−715.

Schal, C. and Cardé, R.T. (1986). Effects of temperature and light on calling in the tiger moth *Holomelina lamae* (Freeman) (Lepidoptera: Arctiidae), *Physiol. Ent.* **11**, 75−87.

Scott, R.W. and Achtemeier, G.L. (1987). Estimating pathways of migratory insects carried in atmospheric winds, *Env. Ent.* **16**(4), 1244−1254.

Sen-Sarma, P.K. (1985). Insect pests of sandal spike disease and possible use of light traps for their control. *In* 'Use of Traps for Pest/vector Research and Control', pp. 103−107, Proc. Nat, Seminar, Mohanpur, West Bengal.

Service, M.W. (1971). Flight periodicities and vertical distribution of *Aedes cantans* (Mg), *Aedes geniculatus* (01), *Anopheles plumbeus* Steph, and *Culex pipiens* L. (Diptera; Culicidae) in southern England, *Bull. Ent. Res.* **60**, 639−651.

Service, M.W. (1974). Further results of catches of *Culicoides* (Diptera: Ceratopogonidae) and mosquitoes from suction traps, *J. Med. Ent.* **8**(3), 283−291.

Service, M.W. (1976). 'Mosquito Ecology: Field Sampling Methods.' Appl. Sci. Publishers, London, 583pp.

Service, M.W. (1977a). Methods for sampling adult Simuliidae, with special reference to the *Simulium damnosum* complex. *Centre for Overseas Pest Res. Tropical Pest Bull.* No. 5, 45pp.

Service, M.W. (1977b). A critical review of procedures for sampling populations of adult mosquitoes, *Bull. Ent. Res.* **67**, 343−382.

Sexton, J.D., Hobbs, J.H., St Jean, Y. and Jacques, J.R. (1986). Comparison of an experimental updraft ultraviolet light trap with the CDC miniature light trap and biting collections in sampling for *Anopheles albimanus* in Haiti, *J. Amer. Mosq. Control Assn.* **2**(2), 168−173.

Shipp, J.L. (1985). Comparison of silhouette, sticky and suction traps with and without dry ice bait for sampling blackflies (Diptera: Simuliidae) in central Alberta, *Canad. Ent.* **117**, 113−117.

Showers, W.B., Reed, G.L. and Oloumi-Sadeghi, H. (1974). Mating studies of female European corn borers: relationship between deposition of egg masses on corn and captures in light trap, *J. Econ. Ent.* **67**, 616−619.

Siddorn, J.W. and Brown, E.S. (1971). A Robinson light trap modified for segregating samples at predetermined time intervals, with notes on the effect of moonlight on the periodicity of catches of insects, *J. Appl. Ecol.* **8**, 69−75.

Silk, P.J. and Kuenen, L.P.S. (1988). Sex pheromones and behavioural biology of the coniferophagous *Choristoneura*, *Ann. Rev. Ent.* **33**, 83−101.

Silk, P.J., Tan, S.H., Wiesner, C.J. *et al.* (1980). Sex pheromone chemistry of the eastern spruce budworm, *Choristoneura fumiferana*. *Env. Ent.* **9**, 640−644.

Silk, P.J., Butterworth, E.W., Kuenen, L.P.S. *et al.* (1989). Identification of sex pheromone component of spruce budmoth *Zeiraphera canadensis*, *J. Chem. Ecol.* **15**, 2435−2440.

Sinsko, M.J. and Craig, G.B. (1979). Dynamics of an isolated population of *Aedes triseriatus* (Diptera: Culicidae). I. Population size, *J. Med. Ent.* **15**, 89−98.

Skovmand, O. and Mourier, H. (1986). Electrocuting light traps for the control of house flies, *J. Appl. Ent.* **102**, 446−455.

Snow, W.F. (1975). The vertical distribution of flying mosquitoes (Diptera: Culicidae) in West African savanna, *Bull. Ent. Res.* **65**, 269−277.

Snow, W.F. (1976). The direction of flight of mosquitoes (Diptera: Culicidae) near the ground in West African savanna in relation to wind direction, in the presence and absence of bait, *Bull. Ent. Res*, **65**, 555−562.

Snow, W.F. (1977). The height and direction of flight of mosquitoes in West African savanna, in relation to wind speed and direction, *Bull. Ent. Res.* **67**, 271−279.

Snow, W.F. (1979). The vertical distribution of flying mosquitoes (Diptera: Culicidae) near an area of irrigated rice-fields in the Gambia, *Bull. Ent. Res.* **69**, 561−571.

Snow, W.F. (1982). Further observations on the vertical distribution of flying mosquitoes (Diptera: Culicidae) in West African savanna, *Bull. Ent. Res.* **72**, 695−708.

Spillman, J.J. (1980). The design of an aircraft mounted net for catching airborne insects. *In*: 'Trends in Airborne Equipment for Agriculture and Other Areas', Pergamon Press, Oxford.

Stanley, B.H., Reissig, W.H., Roelofs, W.L. *et al.* (1987). Timing treatment for apple maggot (Diptera: Tephritidae) control using sticky sphere traps baited with synthetic apple volatile, *J. Econ. Ent.* **80**(5), 1057−1063.

Starratt, A.N. and McLeod, D.G.R. (1976). Influence of pheromone trap age on capture of the European corn borer, *Env. Ent.* **5**, 1008−1010.

Steck, W. and Bailey, B.K. (1978). Pheromone traps for moths; evaluation of cone trap designs and design parameters, *Env. Ent.* **7**(3), 449−455.

Steck, W.F., Underhill, E.W., Bailey, B.K. and Chisholm, M.D. (1982). (Z)-7-tetradecenal, a seasonally dependent sex pheromone of the x-marked cutworm, *Spaelotis clandestina* (Harris) (Lepidoptera: Noctuidae), *Env. Ent.* **11**, 1119−1122.

Steiner, L.F. (1957). Low cost plastic fruit fly trap, *J. Econ. Ent.* **50**, 508−509.

Stinner, R.E., Barfield, C.S., Stimac, J.L. and Dohse, L. (1983). Dispersal and movement of insect pests, *Ann. Rev. Ent.* **28**, 319−335.

Struble, D.L. and Swailes, G.E. (1975). A sex attractant for the clover cutworm, *Scotogramma trifolii* (Rottenberg), a mixture of Z-11-hexadecen-1-ol acetate and Z-11-hexadecen-1-ol, *Environ. Entom.* **4**, 632–636.

Struble, D.L., Swailes, G.E. and Ayre, G.L. (1977). A sex attractant for males of the darksided cutworm *Euxoa meeoria* (Lepidoptera: Noctuidae), *Can. Ent.* **109**, 975–980.

Sudia, J.D. and Chamberlain, R.W. (1962). Battery-operated light trap, an improved model, *Mosq. News.* **22**, 126–129.

Tallamy, D.W., Hansens, E.J. and Denno, R.F. (1976). A comparison of Malaise trapping and serial netting for sampling a horsefly and deerfly community, *Env. Ent.* **5**, 788–792.

Taylor, L.R. (1962). The absolute efficiency of insect suction traps, *Ann. Appl. Biol.* **50**, 405–421.

Taylor, L.R. (1974). Monitoring change in the distribution and abundance of insects, *Rothamsted Expt. Stn Report for 1973, Pt. 2*, 202–239.

Taylor, L.R. (1986). Synoptic dynamics, migration and the Rothamsted Insect Survey, *J. Anim. Ecol.* **55**, 1–38.

Taylor, L.R. and Brown, E.S. (1972). Effects of light trap design and illumination on samples of moths in the Kenya highlands, *Bull. Ent. Res.* **62**, 91–112.

Taylor, L.R., French, R.A., Woiwood, I.P. *et al.* (1981a). Synoptic monitoring for migrant insect pests in Great Britain and Western Europe I. Establishing expected values for species content, population stability and phenology of aphids and moths. Rothamsted Experimental Station, Report for 1980, Part 2, pp. 41–104.

Taylor, L.R., Woiwood, I.P., Tatchell, G.M. *et al.* (1981b). Specific monitoring for migrant insect pests in Great Britain and Western Europe. III. The seasonal distribution of pest aphids and the annual aphid aerofauna over Great Britain 1975–80. Rothamsted Experimental Station Report for 1981, Part 2, pp. 23–35.

Thompson, B.H. (1976a). Studies on the attraction of *Simulium damnosum* s.l. (Diptera: Simuliidae) to its hosts. I. The relative importance of sight, exhaled breath, and smell. *Zeit. Tropenmed. Parasit.* **27**, 455–473.

Thompson, B.H. (1976b). Studies on the attraction of *Simulium damnosum* s.l. (Diptera: Simuliidae) to its hosts. II. The nature of substances on the human skin responsible for attractant olfactory stimuli, *Zeit. Tropenmed. Parasit.* **27**, 83–90.

Thompson, D.V., Capinera, J.L. and Pilcher, S.D. (1987). Comparison of an aerial water-pan pheromone trap with traditional trapping techniques for the European cornborer (Lepidoptera: Pyralidae), *Env. Ent.* **16**, 154–158.

Thompson, P.H. (1969). Collecting methods for Tabanidae, *Ann. Ent. Soc. Amer.* **62**, 50–57.

Thompson, P.H. and Bregg, E.J. (1974). Structural modifications and performance of the modified animal trap and the modified Manitoba trap for collection of Tabanidae (Diptera), *Pro. Ent. Soc. Wash.* **76**(2) 119–122.

Thompson, P.H. and Pechuman, L.L. (1970). Sampling populations of *Tabanus quinquevittatus* about horses in New Jersey, with notes on the identity and ecology, *J. Econ. Ent.* **63**(1), 151–155.

Thorsteinson, A.J., Bracken, G.K. and Hanec, W. (1965). The orientation behaviour of horse-flies and deer-flies (Tabanidae: Diptera). III. The use of traps in the study of orientation of tabanids in the field, *Ent. Expt. et Appl.* **8**, 189–192.

Townes, H.K. (1962). Design for a Malaise trap, *Proc. Ent. Soc. Wash.* **64**, 253–262.

Trueman, D.W. and McIver, S.B. (1981). Detecting fine-scale temporal distributions of biting flies: a new trap design, *Mosquito News.* **41**(3), 439–443.

Tucker, M.R. (1983). Light trap catches of African armyworm moths *Spodoptera exempta* (Lepidoptera: Noctuidae), in relation to rain and wind, *Bull. Ent. Res.* **73**, 315–319.

Tunset, K., Nilssen, A.C. and Andersen, J. (1988). A new trap design for primary attraction of bark beetles and bark weevils (Coleoptera, Scolytidae and Curculionidae), *J. Appl. Ent.* **106**, 266–269.

Turgeon, J.J. and Grant, G.G. (1988). Response of *Zeiraphera canadensis* (Lepidoptera: Tortricidae: Olethreutinae) to candidate sex attractant and factors affecting trap catches, *Env. Ent.* **17**(3), 442–447.

Underhill, E.W., Millar, J.G., Ring, R.A. *et al.* (1987). Use of sex attractant and an inhibitor for monitoring winter moth and Bruce spanworm populations, *J. Chem. Ecol.* **13**, 1319–1330.

Urban, A.J. (1976). Colour vision of the citrus psylla *Trioza erytreae* (Del Guercio) (Homoptera; Psyllidae) in relation to alightment colour preferences. 95pp. Ph.D. thesis, Rhodes Univ. Grahamstown, South Africa.

Vaishampayan, S.N., Krogan, M., Waldbauer, G.P. and Wooley, J.T. (1975). Spectral specific responses in the visual behaviour of the greenhouse whitefly *Trialeurodes vaporariorum* (Homoptera: Aleyrodidae), *Ent. Exp. et Appl.* **18**, 344–356.

Vaishampayan, S.M. (1985a). Factors affecting the light trap catches of insects with emphasis on design aspects. *In* 'Use of Traps for Pest/vector Research and Control', pp. 62–67, Proc. Nat. Seminar, Mohanpur, West Bengal.

Vaishampayan, S.M. (1985b). JNKVV trap SM-model-light trap for survey of insect pests and vectors. *In* 'Use of Traps for Pest/vector Research and Control', pp. 152–155, Proc. Nat. Seminar, Mohanpur, West Bengal.

Vakenti, J.M. and Madsen, J.F. (1976). Codling moth (Lepidoptera: Olethreutidae): monitoring populations in apple orchards with sex pheromone traps, *Canad. Ent.* **108**, 433–438.

Vale, G.A. (1969). Mobile attractants for tsetse flies, *Arnoldia* (Rhodesia), **33**, 1–6.

Vale, G.A. (1974a). Direct observations on the responses of tsetse flies (Diptera: Glossinidae) to hosts, *Bull. Ent. Res.* **64**, 589–594.

Vale, G.A. (1974b). New field methods for studying the responses of tsetse flies (Diptera: Glossinidae) to hosts, *Bull. Ent. Res.* **64**, 199–208.

Vale, G.A. (1974c). The responses of tsetse flies (Diptera: Glossinidae) to mobile and stationary baits, *Bull. Ent. Res.* **64**, 545–588.

Vale, G.A. (1977a). The flight of tsetse flies (Diptera: Glossinidae) to and from a stationary ox, *Bull. Ent. Res.* **67**, 297–303.

Vale, G.A. (1977b). Feeding responses of tsetse flies (Diptera: Glossinidae) to stationary hosts, *Bull. Ent. Res.* **67**, 635–649.

Vale, G.A. (1979). Field responses of tsetse flies (Diptera: Glossinidae) to odours of men, lactic acid and carbon dioxide, *Bull. Ent. Res.* **69**, 459–467.

Vale, G.A. (1980). Field studies on the responses of tsetse flies (Glossinidae) and other Diptera to carbon dioxide, acetone and other chemicals, *Bull. Ent. Res.* **70**, 563–570.

Vale, G.A. (1982). The improvement of traps for tsetse flies (Diptera: Glossinidae), *Bull. Ent. Res.* **72**, 95–106.

Vale, G.A., Flint, S. and Hall, D.R. (1986). The field responses of tsetse flies, *Glossina* spp. (Diptera: Glossinidae) to odours of host residues, *Bull. Ent. Res.* **76**, 685–693.

Vale, G.A. and Hall, D.R. (1985). The role of l-octen-3-ol, acetone and carbon dioxide in the attraction of tsetse flies, *Glossina* spp. (Diptera: Glossinidae) to ox odour, *Bull. Ent. Res.* **75**, 209–217.

Vale, G.A., Lovemore, D.F., Flint, S. and Cockbill, G.F. (1988). Odour baited targets to control tsetse flies, *Glossina* spp. (Diptera: Glossinidae) in Zimbabwe, *Bull. Ent. Res.* **78**, 31–48.

Vogt, W.G. (1986). Influence of weather and time of day on trap catches of bushfly, *Musca vetustissima* Walker (Diptera: Muscidae), *Bull. Ent. Res.* **76**, 359–366.

Vogt, W.G. (1988). Trap catches of *Musca vetustissima* Walker (Diptera: Muscidae) and other arthropods associated with cattle dung in relation to height above ground level, *J. Aust. Ent. Soc.* (in press).

Vogt, W.G. and Havenstein, D.E. (1974). A standardized bait trap for blowfly studies, *J. Aust. Ent. Soc.* **13**, 249–253.

Vogt, W.G., Woodburn, T.L. and Crompton, G.W. (1981). Estimating absolute densities of the bushfly, *Musca vetustissima*, Walker (Diptera: Muscidae) using West Australian blowfly traps, *Bull. Ent. Res.* **71**, 329–337.

Vogt, W.G., Woodburn, T.L., Morton, R. and Ellem, B.A. (1983). The analysis and standardization of trap catches of *Lucilia cuprina* (Wiedemann) (Diptera; Calliphoridae), *Bull. Ent. Res.* **73**, 609–617.

Vogt, W.G., Runko, S. and Starick, M.K. (1985a). A wind oriented fly trap for quantitative sampling of adult *Musca vetustissima* Walker, *J. Aust. Ent. Soc.* **24**, 228–237.

Vogt, W.G., Woodburn, T.L. and Foster, G.S. (1985b). Ecological analysis of field trials conducted to assess the potential of sex-linked translocation strains for genetic control of the Australian sheep blowfly, *Lucilia cuprina* (Wiedemann), *Aust. J. Biol. Sci.* **38**, 259–273.

Vogt, W.G., Woodburn, T.L., Morton, R. and Ellem, B.A. (1985c). The influence of weather and time of day on trap catches of males and females of *Lucilia cuprina* (Wiedemann) (Diptera: Calliphoridae), *Bull. Ent. Res.* **75**, 315–319.

Wall, C. and Perry, J.N. (1978). Interaction between pheromone traps for the pea moth, *Cydia nigricana* (F), *Ent. Expt. et Appl.* **24**, 155–162.

Wall, C. and Perry, J.N. (1980). Effects of spacing and trap number on inter-relations between pea moth pheromone traps, *Ent. Expt. et Appl.* **28**, 313–321.

Wall, C. and Perry, J.N. (1983). Further observations on the responses of male pea moth, *Cydia nigricana*, to vegetation previously exposed to sex attractant, *Ent. Expt. et Appl.* **33**, 112–116.

Wall, C. and Perry, J.N. (1987). Range of action of moth sex-attractant sources, *Ent. Expt. et Appl.* **44**, 5–14.

Webb, R.E., Smith, F.F., Affeld, H. *et al.* (1985). Trapping greenhouse whitefly with coloured surfaces; variables affecting efficacy, *Crop. Protect.* **4**(3), 381–393.

Wardhaugh, K.G., Read, P. and Meave, M. (1984). A sticky trap for studying the spatial distribution of the Australian sheep blowfly *Lucilia cuprina*, *Aust. Vet. J.* **61**, 132.

Wardhaugh, K.G., Smith, P.H. and Crompton, G.W. (1983). A comparison of ground and aerial release methods for the genetic control of *Lucilia cuprina* (Wiedemann) (Diptera: Calliphoridae), *Gen. App. Ent.* **15**, 37–46.

Warnes, M.L. and Finlayson, L.H. (1985a). Responses of the stable fly, *Stomoxys calcitrans* (L.) (Diptera: Muscidae), to carbon dioxide and host odours, I. Activation, *Bull. Ent. Res.* **75**, 519–527.

Warnes, M.L. and Finlayson, L.H. (1985b). Responses of the stable fly, *Stomoxys calcitrans* (L.) (Diptera; Muscidae), to carbon dioxide and host odours, II. Orientation, *Bull. Ent. Res.* **75**, 717–727.

Weatherston, J., Roelofs, W., Comeau, A. and Sanders, C.J. (1971). Studies of physiologically active arthropod secretions. X. Sex pheromone of the eastern spruce budworm, *Choristoneura fumiferana* (Lepidoptera, Tortricidae), *Can. Ent.* **103**, 1741–1747.

Welch, J.B. (1988). Effect of trap placement on detection of *Cochliomyia hominivorax* (Diptera: Calliphoridae), *J. Econ. Ent.* **81**, 241–245.

Wenk, P. and Schlorer, G. (1963). Wirtsorientierung und Kopulation bei blut-saugenden Simuliiden (Diptera), *Zeit. trop. Med. Parasit.* **14**, 177–191.

Weslien, J and Bylund, H. (1988). The number and sex of spruce bark beetles, *Ips typographus* L. caught in pheromone traps as related to flight season, trap type, and pheromone release, *J. Appl. Ent.* **106**, 488–493.

Westigard, P.H. and Graves, K.L. (1976). Evaluation of pheromone-baited traps in a pest management programme on pears for codling moth control, *Canad. Ent.* **108**, 379–382.

Williams, C.B. (1936). The influence of moonlight on the activity of certain nocturnal insects, particularly of the family Noctuidae, as indicated by a light trap, *Phil. Trans. R. Soc. (B)* **226**, 357–389.

Williams, C.B. (1962). Studies on black-flies (Diptera: Simuliidae) taken in a light trap in Scotland. Part 3. The relation of night activity and abundance to weather conditions, *Trans. R. Ent. Soc. Lond.* **114**, 28–47.

Williams, C.B. (1965). Black-flies (Diptera: Simuliidae) in a suction trap in the central highlands of Scotland. *Proc. Roy. Ent. Soc. Lond.* **40**, 92–95.

Williams, D.F. (1974). Sticky traps for sampling populations of *Stomoxys calcitrans*, *J. Econ. Ent.* **66**(6), 1279–1280.

Wilton, D.P. and Fay, R.W. (1972). Air flow direction and velocity in light trap design, *Ent. Exp. et Appl.* **15**, 377–386.

Woiwood, I.P. and Dancy, K.J. (1986). Synoptic monitoring for migrant insect pests in Great Britain and Western Europe. VII. Annual population fluctuations of macrolepidoptera over Great Britain for 17 years, *Ann. Rev. Rothamsted Expt. Stn.* 237–238.

Woiwood, I.R., Tatchell, G.M. and Barrett, A.M. (1984). A system for the rapid collection, analysis and dissemination of aphid monitoring data from suction traps, *Crop. Protect.* **3**, 273–288.

Woiwood, I.P., Tatchell, G.M., Dupuch, M.J. *et al.* (1985). 'Rothamsted Insect Survey', 17th annual summary, pp. 253–255.

Woiwood, I.P., Tatchell, G.M., Dupuch, M.J. 'Rothamsted Insect Survey', 18th annual summary, pp. 289–293.

Wyman, J.A. (1979). Effect of trap design and sex attractant release rates on tomato pinworm catches, *J. Econ. Ent.* **72**, 865–868.

Zaim, Z., Ershadi, M.R.I., Manouchehri, A.V. and Hamdi, M.R. (1986). The use of CDC light traps and other procedures for sampling malaria vectors in southern Iran, *J. Amer. Mosq. Control Assn.* **2**(4), 511–515.

Zervas, G.A. (1982). A new long-life trap for olive fruit fly, *Dacus oleae* (Gmelin) (Dipt., Tephritidae) and other Diptera, *Z. ang. Ent.* **94**, 522–529.

Index

A- COOK